Believers – March 13, 2016   BH

A Gift to You
from
The Downtown
Jewish Women's
Connection

Sponsored by:
The Brandon Merritt
Charitable Foundation
www.bthedifference.org

Marla Bergmann &
Michael Melamud

Marcy & David Kronrad in
memory of Marvin Hyman

Lys & Bill Rubin

Stacey & Benny Shabtai

Downtown Jewish Center Chabad
Bet Ovadia Levy

37 Seconds

37 Seconds

DYING REVEALED HEAVEN'S HELP—
A MOTHER'S JOURNEY

Stephanie Arnold

with Sari Padorr

HarperOne
An Imprint of HarperCollinsPublishers

HarperCollins books may be purchased for educational, business, or sales promotional use. For information please e-mail the Special Markets Department at SPsales@harpercollins.com.

HarperCollins website: http://www.harpercollins.com

HarperCollins®, 🏠®, and HarperOne™ are trademarks of HarperCollins Publishers.

FIRST EDITION

Designed by Terry McGrath

Library of Congress Cataloging-in-Publication Data
Arnold, Stephanie.
37 seconds : dying revealed heaven's help—a mother's journey / Stephanie Arnold, with Sari Padorr. — First edition.
    pages cm
ISBN 978–0–06–240218–9
1. Near-death experiences. 2. Reincarnation therapy. 3. Post-traumatic stress disorder—Alternative treatment. 4. Pregnancy—Complications—Miscellanea.
5. Cesarean section—Complications—Miscellanea. I. Title. II. Title: Thirty-seven seconds.
BF1045.N4A76 2015
133.901'3—dc23

2015010918

15 16 17 18 19   RRD(H)   10 9 8 7 6 5 4 3 2

To my children, Valentina, Adina, and Jacob.
Thank you for loving me with the purest, strongest, and
most sensitive of hearts. Everything I do, I do for you.

To the love of my life. I will never forget the pain
we endured or the strength it took to get to the other side.
I love you, sweetheart, in this lifetime and the next.

The first time is a charm: Jonathan and I during our first pregnancy together, which now strikes us as amazingly uneventful.

Prologue

I'VE ALWAYS KNOWN YOU TO BE A FIGHTER." It was a pep talk just like the ones he had given me many times before. "I've always known you to survive," my Uncle Marvin said, trying to comfort me. I started to sob uncontrollably. And then I realized he was trying to distract me from turning around. Suddenly, a sharp pain shot down my throat and my stomach felt like it was being ripped open. I turned around and watched in horror as a scalpel dug deep into the sternum of someone on an operating table, slicing all the way down the center of the person's stomach. When one of the doctors moved to the side, I saw that the someone was me.

"Can I come out?" I gasped. I wanted out of hypnosis. "I think I'm going to throw up."

"I'm going to count . . . one, two, three, four, five. Take another deep breath, you're coming back. It's all okay," Linda, my therapist, softly said. Tears were streaming down my face.

Did I just see what I thought I saw? Was I actually witnessing doctors trying to bring me back to life? I was in shock. For months people had been asking me what I could remember about the moment I died on the operating table, but I couldn't remember a thing. Now, during a regression therapy session, I was watching and feeling every painful detail. Did it really happen that way? Was I looking through a window into the past? Or was this a recalled episode of *Grey's Anatomy* left in my subconscious? I couldn't get my head around it.

Eight months before, I died after giving birth to our son, Jacob. I flatlined for 37 seconds. What happened to me was medically unpredictable, but my doctors will tell you that I survived because I predicted it. I had experienced detailed premonitions for months beforehand that I would die the same day my son was born, and many—including my doctors—believe those visions saved my life.

A few months after Jacob's birth—after I had recovered from not just childbirth but trauma and death—I realized I needed help processing the entire experience. The realization that I had seen my own death ahead of time was too much to handle on my own. My doctors couldn't give me any direction, religious leaders were sure that G-d (which is how many of us in the Jewish community spell the term in order not to inadvertently take the name in vain) had played a hand in it, and my husband was happy that I was alive but, as a left-brain thinker, couldn't even begin to fathom how I could have seen my own death. I needed more help than my doctors and rabbis could offer. So, I turned to regression therapy.

No one prepared me for the pain I was about to endure by going back into the past or for what I was about to see.

And if it was possible to go back into the past, then was it possible that I actually had a conversation with my uncle? After all, he had been dead for more than 20 years.

I would soon discover, as amazing as this is going to sound, that the details I saw by looking into the past were as accurate as my premonitions had been. These experiences ultimately opened up doors to a world I never knew existed and certainly never thought I would someday see. This is my story, as it happened to me and through the eyes of everyone else who witnessed it.

Chapter 1

S HE READS STAINS. Coffee stains at the bottom of a very
strong cup of Turkish coffee. I hate coffee, but I guess
my curiosity got the best of me, so when I was 19 years old I
went to see this psychic and drink from the cup. I wanted to
see what the future would bring. Was it a boatload of riches,
a handsome man, or a job that made me a great success? I
never expected to hear what she told me. She said I would
"die at an early age." I know, right? They never tell you bad
news, but there it was. No wealth. No man. No, I was going
to die at an early age.

Of course, back then I passed off that prediction as maybe just a psychic's ploy to get more money out of me by forcing me to ask how and when. I didn't bite. Maybe I should have.

I started thinking about that reading as I was recovering from dying.

Twenty-two years after the coffee lady allegedly saw those fatal stains, I died at the age of 41 for 37 seconds. Did the psychic really see something back then, or was it just coincidence? I'll never know. But I have come to believe that I can't discount the possibility of being able to see into the future, because months before I was to give birth to our second child, I had visions that I was going to die. They were scary and detailed. At the time I couldn't have told you why or how these visions happened, but they would ultimately save my life. In the aftermath, as I relived my death—only this time as an observer—I came to understand who had sent me those warnings.

It was May 30, 2013, one week shy of my scheduled C-section, and I woke up with a craving for a cigarette. I don't smoke. Never have. But throughout my pregnancy, I had craved them constantly. It was weird. I certainly wasn't going to pick up a smoke, but I had found myself purposely walking close to smokers just to get a whiff. Strange, I know.

I shook off the craving and headed to the kitchen to start the day. I was giving my daughter, Adina, breakfast, and I felt off. All of a sudden I felt a strange cramp, looked into my underwear, and saw blood. It wasn't a few droplets but a full rush of blood that quickly soaked my nightgown and puddled at my feet. Adina stood there staring at me with fear in her eyes. The mommy in me went into high gear—as I tried to clean up. The only thing I could think to do was to keep smiling, be positive, and tell her she was going to meet her brother today. I didn't want her to be scared, even if I was.

The truth is, I was terrified. I wasn't in pain, just in shock. And from the look on my daughter's face, she felt the same way. So I put a smile on my face and got excited, and she was distracted enough not to focus on the fact that I was, in her words, "peeing red."

I calmly called Tessie to come upstairs. She was Adina's night nurse when she was born. Tessie was also a close family friend, and she had told my husband that she wouldn't leave me alone while he was out of town, given the pregnancy complications I was having. I am so grateful she was there that morning.

I knew I needed help right away, so I prepared to head to the hospital. As we got Adina strapped into the car, a million things were going through my mind.

I had to call my husband, Jonathan, who was in a meeting in New York. I could have walked to the hospital located directly behind our house in Chicago to get immediate medical attention, but I wanted to be with my doctors, who knew my medical history. So I knew I had to drive. Tessie went to get into the driver's seat, but I snapped at her and told her to get out of the way. I knew she would be nervous and wouldn't know where to go. Giving her directions and telling her to run yellow lights would only make her panic, so I told her to get into the backseat. I was still bleeding, but I figured I could get to the hospital in 15 minutes flat. I felt I had enough time. Okay, maybe it wasn't the smartest move I've ever made, and yes, I know what could have happened. Still, I'd had many fears and premonitions leading up to that day, but dying in a car accident wasn't one of them.

I started to pray. I remember my family rabbi teaching me at an early age that the Shema prayer means many things, but that its main purpose is to protect. Jews are supposed to say this prayer every morning and every evening so that its power will help to protect their souls. I desperately needed that now. And I was certain that if no one else could help me, my Jewish faith and G-d could. So I prayed.

I was in control and driving as calmly as I could. The blood was contained, for the moment, with a pad. I called my husband and told him he needed to get on a plane to Chicago because we were having our baby that day. I called my father to tell him I was on my way to the hospital, but I didn't share my fears with my parents, as it would have done nothing but scare them. I had this sinking feeling that I was a ticking time bomb. For the sake of my daughter, I compartmentalized my fear and soldiered on.

In the short time it took that May day to rush to Prentice Women's Hospital at Northwestern Medical Center in Chicago, I realized that this moment was the culmination of all my hopes and fears: my hope to expand my family with the man I so deeply loved, and my fear that I wouldn't make it through the delivery. It was more than just a normal fear. I knew I was going to die.

For the first five months, my pregnancy had been physically perfect—the complete opposite of what I had experienced with our first child, Adina. This time there was no morning sickness, no acid reflux or leg cramps. Nothing.

I spent much of my second trimester getting our home in Chicago ready for sale so that we could move to New York. My husband, Jonathan, had accepted a position with the New York Attorney General's Office as its chief economist. It was a dream job for him. He'd wanted to go into government service for as long as I'd known him. But it was a decision we didn't take lightly, since we were planning to expand our family and let go of the roots we had planted in the Windy City. Flying between Chicago and New York was becoming a weekly ritual for us. I made sure to have obstetricians lined up in both cities, just to play it safe.

It was February 7, and I had just flown back to New York in time for the all-important 20-week ultrasound. This is the screening where they look at the spine and all of the organs and can see more clearly if there are any serious complications with the baby. I was thinking about my easy pregnancy when the radiologist poked his head in after the test and said, very bluntly, "You have a complete placenta previa. I'll be right back," then rushed out of the room to take a call.

What? *What!?!* What does that mean? Is the baby in danger? Am I in danger? I turned to Jonathan and said, very determinedly, "I have a rare blood type, and I don't know what a complete placenta previa is, but I have a bad feeling about this."

"Let's not get ahead of ourselves," Jonathan said, calmly and rationally. "Let's learn about it and take the appropriate precautions."

When the radiologist came back in, he said, "It's nothing to worry about. As the uterus grows it might move out of place. You don't have to restrict your activity. It's no big deal."

That was it. Nothing more. Jonathan was relieved, but I was already dialing my OB/GYN in Chicago, Dr. Julie Levitt, as I was walking out the door. Her response? "Absolutely you need to restrict your activity. I just had a previa that had to stay on hospital bed rest for 72 days before giving birth at 35 weeks." *It is a big deal,* I told myself.

After getting our family fed and settled in the apartment that cold February night, I started Googling "complete placenta previa." It's a condition where the placenta "lies on the bottom of the uterus blocking the cervix." My heart started racing. I read on. "It can cause bleeding. And possible complications. And requires a C-section."

Chills ran down my spine. Complications?! I'd already had a C-section when Adina was born, and the likelihood of having a previa after one C-section was less than 5 percent. Still, my head was reeling. I read screen after screen about all the complications, wanting to take in as much as I could. Then, as if I were watching a movie on my computer screen, I saw a flash and my future unfolded.

I saw myself on the operating table. I saw the doctors working feverishly on me. I saw Jonathan holding our newborn Jacob, who was fine. But I was not. I saw my mouth open and my body heavily placed on the operating table like a slab of meat. I was dead.

What was that?! I shook my head in disbelief. *What did I just see?* I had no idea where those images came from. I started panicking. I had to catch my breath. How could I see something like that? Why would I see something like that? Was I so freaked out that I was causing my mind to imagine the worst-case scenario?

I ran into the bedroom and told Jonathan about it. Once again, he tried to calm me down, saying, "That is the absolute worst-case scenario. There's no need to panic. Nothing's going to happen."

But I knew this wasn't just my imagination. I had a visceral reaction to what I saw. My hands were freezing cold down to my fingertips. My body felt extremely heavy. This wasn't normal pregnancy anxiety. I already had given birth once before, and this feeling was nothing like the jitters of my first delivery. I wasn't sure why I was seeing it, but I immediately felt that the "vision" was real. I was convinced that I would not survive the birth of my boy.

Chapter 2

THE DAY AFTER THAT FIRST VISION, I had another one. This time I was pushing Adina in the stroller across a park in Tribeca on the way to her school. I was still shaken from what I'd seen the night before, which I was playing over and over in my mind, trying to figure out what the heck it was. I was concerned about "seeing" this mini-movie, but more concerned about what I saw. I had never hallucinated, suffered a nervous breakdown, or imagined things that weren't real.

I kept telling myself to calm down and take deep breaths.

Adina was excited to get to class, and her small voice and tiny smile helped to bring me back from my thoughts.

As I walked through the brisk air I saw the empty fountains and imagined how pretty they would be once the water was turned back on in the spring. I could almost picture the flow of the water, the fountains spraying in every direction. That's when the next vision struck. All of a sudden that water flow I was imagining turned into blood. But I wasn't looking at a fountain—I was seeing inside myself as blood started to ooze and pool and hemorrhage. I saw blood coming out of my veins and pouring out over my uterus and running down the insides of my legs. And I not only saw it—I felt it.

As quickly as the vision came into my mind's eye, it was gone. I held on to the stroller for balance, to keep myself from falling down. The sounds of traffic snapped me out of it. *What the hell was that?* It had looked real, and worse than that, it had felt real. *Am I going crazy? Is this what a nervous breakdown looks like?*

Adina and I quickly arrived at the mom-and-tot class. I went through the motions, but I wasn't mentally there, and as soon as we got out of class I called Jonathan. This time he seemed a little more concerned about what I was telling him. Once again, he tried to console me without really understanding. But how could he? I didn't really understand what was going on. I did know, however, that whatever it was, it wasn't good.

I did everything in my power to keep my fear at bay and do my normal routine.

A few days later, I was doing the weekly grocery shopping, not paying much attention to what I needed as I wandered aimlessly down each aisle. I turned the corner to head into the baked goods section, and I saw the next one clear as day. It was another mini-movie, and this one rocked me to my core. I saw my funeral.

I was inside a shiny, coffee-colored casket and wearing Jonathan's Air Force dog tags. He was wearing a religious shroud as he kneeled next to the coffin and cried. There was a white awning over the burial site, which sat on a small hill overlooking hundreds of headstones and water.

The next image was Jonathan saying a prayer. It's called the Eshet Chayil, or the "Woman of Valor" prayer from Proverbs 31, and he says it to me every Friday night. It's his way of blessing me for taking care of the family and also letting me know how he feels about being married to me. I watched our family starting to make challah, the kids kneading the dough as they put more flour on the floor than in the bowl. Trying to keep his composure, Jonathan was wiping the tears away from his eyes so the children couldn't see how terribly sad he was. It was clear that he was being protective of them, praying for their joy and peace and trying as best he could to prevent them from feeling the pain that would eventually surface from my absence.

Once again, I wasn't imagining it—I was seeing it.

What was happening to me? Was I going crazy? Or were these visions as real as they felt?

UNTIL THEN, I had always been very calm and collected. I had to be that way to work in the television industry. Over the years I had produced television programs and music videos with stars like Sarah Jessica Parker and Julio Iglesias, and the crazy, fast-paced world of production would have eaten me up if I hadn't been in control.

Throughout my youth and as a young adult, I was driven to get ahead and was married to my career. Relationships always came second to my work, and having a family was never a consideration. It wasn't as if I didn't date or believe in love. I took a brief detour in my twenties and got married for all the wrong reasons. It didn't last, and from that point on I never looked into the future any further than my next deadline. I was convinced that marriage wasn't for me. But that all changed the day I met Jonathan Arnold.

I was living in Los Angeles, and a mutual friend suggested that I meet "this economist who is really great." I didn't even know what an economist actually did, let alone whether I would have anything in common with him. Jonathan lived in Chicago but was in LA frequently to visit his daughter, Valentina. He called me on one of his trips to LA, and what transpired over the next 45 minutes was a phone call unlike anything either of us had ever experienced. We discussed religion, past relationships, politics, and even where we saw ourselves in five to ten years.

It was as if I had known him my entire life. It's hard to put it into words, but it was like an instant magnetic force drawing us to one another. Jonathan said later that he hung up and thought, *I think I just spoke to the woman with whom I'll spend the rest of my life.* I felt the same way, but for me it would take a first date to make sure the energy I was feeling over the phone was real. When I saw him four days later, I was completely convinced and did a total 180. I wanted to marry him.

We saw each other night after night for the next three days and spoke every chance we could. On day eight of this whirlwind courtship, he wrote me a love letter and posed a few questions:

> *How can a man go from zero to love in three dates—*
> *even with one as spectacular as you?*
> *What gives me the ability to catch you—the girl who*
> *specializes [in] avoiding capture?*

I wished I could be as poetic, but the only thing I could think of to do in return was make him a mixtape. I felt like I was back in high school.

We were both smitten. So much so that I couldn't marry

him quickly enough. A few months after that first phone call, we decided to elope. We wanted the special day to be just between us and Valentina. Everything happened so quickly, and Jonathan ordered $40 wedding bands on the Internet for next-day delivery. We then planned another wedding for our family and friends several weeks later, with a second set of "proper" wedding bands.

I watched my career fade away in the rearview mirror as I moved to Chicago, and believe it or not, I didn't take along one regret. I settled into my new "June Cleaver" life, and everything that I had once thought important became irrelevant instantly. I went from not wanting a family to wanting to bear *his* children. Zero to 100 in a matter of minutes.

The love between us grew more intense by the day. He would frequently write me love letters, and his beautiful prose left me speechless:

Victor Hugo once wrote, "Life is a flower for which love is the honey," and David Viscott wrote, "To love and be loved is to feel the sun from both sides."

These words, beautiful and true, fail to begin to capture the overwhelming nature of my feelings for you . . . and for our love.

You are pulchritude personified physically, emotionally and spiritually—and I am therefore envious of me.

His words made me deliriously happy. I had found my soul mate, and I was looking forward to sharing a full life and growing old with him.

BUT NOW, pregnant with our second child together, my fear and the baby were the only things growing. I thoroughly believed that I was going to die giving birth to our son and be taken away from my children and the man I loved so deeply. It was too agonizing to bear.

The "premonitions," as I began referring to them, kept coming. One showed me, as I watched my organs merge, that I would need a hysterectomy. I saw the uterus and the placenta melding into one another. It looked like two blobs in a red-hot lava lamp fusing together—the blood and cells and tissue and organs converging. It was terrifying. During all of these visions, I was sure that Jacob would be fine. I was also sure that I would not be.

Jonathan, being the good loving husband, tried to calm me down, but for once his words had no impact on me. I spent hours on the Internet searching for any answers to what I was seeing. The "melding" condition I was visualizing— where the placenta merges with the uterus—is called placenta accreta. "Having an accreta that would lead to a hysterectomy is such a rare event, less than 1 percent. Try to put it out of your head," he would say, trying to comfort me.

Emotionally, I couldn't calm down. Logically, I understood exactly where he was coming from.

Jonathan finds comfort in numbers. As the child of a Foreign Service officer, he grew up moving from country to country. That made it difficult for him to make friends and have deep connections to people, so he turned to books.

Math became his first love and would continue to be front and center as he got his undergraduate degree, MBA, and PhD in economics at the University of Chicago. Jonathan served in the Air Force as an officer flying supersonic military jets, and through emergency drills and mock disasters, he learned how to stay calm in a crisis.

This was a crisis. He calculated the probable outcome and the worst-case scenario. While he was doing his best to calm me in his own way, his calculations gave me little comfort. Jonathan was using every tool in his toolbox to try to alleviate the anxiety I was feeling, and nothing was working. He was frustrated that nothing he did could comfort me and that nothing I said added up to anything I should worry about in his mathematical mind. He thought the hormones were getting the best of me. But I was sure that my premonitions couldn't be chalked up to estrogen. I was experiencing a *knowing*—the same sort of knowing I'd had that Jonathan was my soul mate.

The consistency of those images put me into full panic mode. Every time I looked at Adina, I would cry. I felt helpless and hopeless and found myself talking about my incredible fears with friends and even people I didn't know well. My best friend, Rosalind, who was always the voice of reason and had a very strong sense of faith, said, "Don't get crazy. It's in G-d's hands. You will be fine." I told my cousin Sari, who reassured me that "nothing is going to happen. It will all be okay." I told the wife of my husband's friend that I was going to die. I just blurted it out. I can still remember her expression, as we didn't know each other very well. She was a psychologist who specialized in pregnancy issues, couples therapy, and fertility. She was pregnant at the time too, and her eyes were wide open. She had told her husband, she told me later, that she had never heard a pregnant woman speak that way. She was scared for me, but could do little to help.

I was desperately looking for someone to throw me a lifesaver, but it wasn't happening. My trainer quickly changed the subject when I told him, and the woman next door explained how she had pregnancy nightmares with her second child and that was probably what was happening to me. Ahhhhhhhh! That was not it at *all!* I was beyond frustrated and scared out of my mind that either everything was going to come true or I was really losing it. I was hoping for the latter, but somehow I knew the former was in the cards for me. No one was taking me seriously. This wasn't fear of delivering! This was fear of dying!

I decided I needed divine intervention. I reached out

to my rabbi, with whom I had grown close when I lived in Los Angeles. Rabbi Chaim Mentz is an Orthodox Jew, and when we met we became instant family. I met him through a mutual friend, and I immediately felt a lot of care and compassion from him and his family. He is my go-to guru on all things Judaic and, for that matter, on many life decisions. When you're with him, he exudes an assurance that nothing bad will happen. I called him, and he knew immediately what I needed to do. He said, "You need to take Jonathan to the burial site of Rabbi Menachem Schneerson. It is the role of the husband and father to pray for the family."

Rabbi Schneerson, also known as "the Rebbe," was considered one of the most influential Jewish leaders of the twentieth century. He was the head of the Chabad-Lubavitch movement, and they say that everyone he met was always deeply changed by his presence. I had been lucky enough to meet him two years before he died. On certain days of the year, people would line up to meet him in Brooklyn. He would hand out a dollar to each person, and as you waited with bated breath, he would give you some profound insight. You could ask for a blessing and even tell him your worries, and he would always have something incredibly uplifting to say. I was there on one of those days, and he handed me a dollar. I was 21 years old, without a care in the world and no prayer. I didn't even know why I was wasting his dollar, but he said to me, apropos of nothing, "You will have children someday, but it will be a difficult road." That was it. I left his presence, and the next day, as an ignorant, naive teenager defiantly would do, I spent the dollar on a Coke and never thought about it again. That is, until after my son's birth.

Many Hasidic Jews believe that the Rebbe is the Messiah; even though he has passed away, they believe he still emanates a power that will answer their prayers. I don't know whether I believed all of that, but I was willing to try anything. Maybe Jonathan would experience some sort of energy at the burial site and come to realize that what I was feeling wasn't hormonal. Maybe there would be a sign and I would get relief from knowing I would not die. At least, that was what I was hoping.

I'd always had a blind faith in G-d, but not necessarily in the interpreters of religion. I went to a religious school when I was young and learned a great deal about Judaism. But instead of giving me answers, this religious education just prompted more questions. I guess I found it hard to accept some aspects of religion because they had to be accepted on faith, and it was hard for me to embrace things I couldn't see with my own eyes. But I did believe in G-d, and I had, on occasion, asked for his help. This would be one of those times.

Rabbi Mentz said he would fly to New York and take Jonathan and me to the Ohel—the building at Montefiore Cemetery in Queens where the Rebbe was buried. Some people call it the "American Western Wall." There's a place there to sit and write out your prayer. Jonathan and I sat separately to think about what we wanted to write. I knew that you weren't supposed to say a prayer for yourself, for your own health, happiness, or well-being. I also knew, without thinking about it very long, exactly what I was going to write: that Jacob would be a happy, healthy boy. I was hoping, almost praying, that Jonathan would write a prayer for me, but I couldn't tell him to do that. It had to come from his own spiritual place.

When we were done with writing down our prayers, we walked through the cemetery to the smaller building, which consisted of only one room with two doors—one entrance for women, the other for men. A wall of candles and prayer books adorned the dim hall. We went through our respective doors and met up again on the other side.

Rabbi Mentz told us to softly read our prayers out loud toward the Rebbe's grave and then rip up the pieces of paper the prayers were written on and throw them on the grave. We didn't hear what each other said, but I would learn later that our prayers were exactly the same: "That Jacob will be a happy, healthy boy." I also learned there wasn't a prayer for me. Jonathan would tell me that he never, ever thought that anything would happen to me and that he didn't want a negative thought to pass over his lips for fear it would affect my outcome. Did that mean he was more spiritual than he cared to admit? We would discuss that later. But at the cemetery he had the passing thought that maybe the pregnancy was tumultuous because there was something wrong with the baby, so that was where he decided to put his energy.

To tell you the truth, what I thought would be an incredibly inspirational moment for me turned out to be incredibly sad. I was devastated that the positive energy vibes were not being sent to me. We were at a holy burial ground where prayers are supposedly answered. If there was to be any divine intervention, this would be the place to get it. But with no prayer uttered in my name, it didn't seem as though that would be happening.

If my faith, family, and friends and the rabbis couldn't help me, maybe doctors could. I told Jonathan that I wanted to make appointments with as many doctors as I could. He said he was concerned that if we went to see too many doctors, they would do tests that would harm the baby.

It was a good point. G-d forbid that anything would happen to Jacob. We had worked so hard to have our children. We weren't able to get pregnant the normal way, so I went through seven rounds of in vitro fertilization to get our babies. It was a painful process that took us through several failures and kept us on an emotional roller coaster.

I most certainly didn't want anything to happen to this precious cargo I was carrying, but I was at my wit's end. I didn't want my thoughts to propel me into any action that would harm the baby. I had gone on pregnancy message boards to learn about panic attacks and premonitions, thinking they were one and the same, but no one's experience had come close to what was happening to me, so I stopped looking online. I decided to deal with my problem offline.

I did only what the doctors would allow me to do at this late stage in the pregnancy. I did blood work to check all of my hormone levels. When I started lightly spotting at one point, I raced to the hospital to have an ultrasound. It all looked okay, but I still insisted on having ultrasounds as often as I could. I wanted to have that visual contact with the baby, but that wasn't my main motivation for having them done. I was thinking that the ultrasounds might pick up the problem I was foreseeing, finally validating all of my fears, and that a plan of action would be put in place. Unfortunately for me—and fortunately for Jacob—the ultrasounds only saw what they were supposed to see: a healthy, growing boy.

During one of my ultrasounds I thought back to when I had hemorrhaged at 10 weeks. They tell you not to worry if the blood is black. It's probably just old blood being cleaned out after implantation. But at 10 weeks I had started to bleed bright red blood, not black. A lot of it. Interestingly enough, I didn't think we were losing the baby. I felt I was losing something else inside of me. I know it doesn't make a lot of sense, but it was a feeling that something was wrong, just not with the baby. The doctors told me at the time that I was probably bleeding because there had been a pocket of air or blood underneath when the placenta attached itself to the uterus and it had to be expelled. Sort of like when you put wallpaper on a wall. If there's a pocket of air, the paper won't lie flat until you smooth it out. The placenta needed a smooth surface. I wasn't comfortable with that explanation, but the baby was fine so that was all that mattered back then.

ONCE THE PREMONITIONS were in full swing, Jonathan and I decided we needed a distraction to get our minds off of everything. On March 30, when I was about two months away from my delivery date, we went to the opera. My husband was a regular. He knew all of the shows and who was starring in them, but I was a neophyte. I just wanted to be swept away by great music and amazing performances and enjoy a night out without being distracted by my fear. Placido Domingo was performing in Verdi's *La Traviata* (*The Fallen Woman*) at the Met. A friend of Jonathan's had given us her season tickets to that sold-out performance, and lucky us, the seats were front row, center. Crap. Great seats for a great performance, but not for a woman who was seven months pregnant with a baby bearing down on her bladder. I can tell you I was beyond nervous. As we walked down the aisle to our seats, I stopped and went back to the bathroom. I walked back down the aisle again, and before I got to our row I went back out to the bathroom. I needed my bladder to make it through to intermission.

The opera began. Violetta is the main character of this three-act show. She's a prostitute and in love with the royal Alfredo. Alfredo's father, Giorgio, played by Placido Domingo, comes to her and tells her she needs to leave Alfredo because her reputation will disgrace his family if she continues their relationship. She reluctantly agrees, for the sake of her love. Alfredo is in love with her and will do anything for her. What neither man knows is that Violetta is dying. She forces Alfredo to leave her by telling him she is not in love with him when, in reality, she wants him to love her until her last breath. Giorgio finally confesses to Alfredo that he was the cause of their separation. Violetta and Alfredo reunite with great passion—and then she drops dead.

It was then that I realized this was probably not a good show to see when you're thinking about your own impending death.

I started to cry, and not just watery eyes because of the moving performances. I sobbed with a sorrow-filled heartache. I was sniffing and shaking. As I got louder and louder, some of the orchestra members looked up at me, probably thinking they were getting a great reaction from a deeply moved audience member. But I was thinking about my own mortality. I was thinking that in a few short weeks my life would be over and my husband would be left with pain too severe to bear. I was thinking about my daughter and wondering if she would remember me. I was thinking about my stepdaughter Valentina—would she remember the amazing memories we had created? But mostly I was thinking about the love I had searched for my entire life and only found a few years before. I was not ready to give it up. Jonathan looked at me and knew instantly what I was thinking.

A couple of days later, on April 1, I was back in Chicago when I felt a cramp and started to bleed.

Chapter 3

"Only go up and down one flight of stairs a day. No picking up Adina or lifting heavy things. And no driving," ordered Dr. Julie Levitt, my Chicago obstetrician. Although she said everything looked okay, she wanted to cut down the chance that I would bleed again and need bed rest and hospitalization. I listened and obeyed, but it wasn't easy. I was in Chicago with Adina, and Jonathan was working in New York, so when we got this news we decided to call for help. When Tessie agreed to move in through the birth of our son, that relieved a lot of the stress. Jonathan rearranged his schedule to be able to come to Chicago on the weekends, and we decided to put the sale of our home on hold until after the baby was born.

I was taking life slowly, but my mind kept racing. I was still consumed by the premonitions. I couldn't sleep. I would lie awake at night just ruminating. I felt like I was running out of time. My internal conflict between wanting to keep trying to find someone who could help and being paralyzed about what to do next was growing.

I sucked up my fear and made a move. I contacted a friend of Jonathan's who was a gynecologic oncologist and told him I was afraid of needing a hysterectomy. He told me that it wasn't going to happen, but that if it did, I would want a gynecologic oncologist to do the procedure because that specialist focuses on highly complex female surgeries. A hysterectomy immediately following childbirth would be included in that category. He referred me to Dr. Julian Schink, the head of gynecologic oncology at Northwestern Memorial Hospital.

When I called Dr. Schink's office, the receptionist said it was unusual to request to see the doctor about the possibility of a hysterectomy when there was no evidence I would need one. After all, he focused on patients with cancer of the reproductive organs. She kept asking me, "What kind of cancer do you have?" and "Are you sure you want to see Dr. Schink?"

"Yes," I said. "I am seven months pregnant, but do not have cancer. I just feel I may need to have a hysterectomy."

She thought I was nuts. I was getting increasingly anxious the more we talked. After going back and forth several times while she tried to figure out how to categorize the visit, I hit my breaking point. Loudly I said, "Can I just make the appointment?" She finally did.

On April 15, Jonathan flew in and we met Dr. Schink. As we waited in the lobby to be called in I scanned the room and saw the faces of women stricken with cancer. They looked gray and sick, with no eyebrows or hair on their heads. Some were carting around IV poles. I didn't want to make eye contact because I was staring at mortality and I was an otherwise healthy, cancer-free pregnant woman. I felt uncomfortable being there. Jonathan said he was embarrassed to be there taking up this doctor's time when everyone else was in a life-or-death situation. *But aren't I?* I thought.

Of course I felt a little hesitant to see this doctor, knowing that he would look at me like I was a hysterical, hormonal mess of a woman, but by that point I was accustomed to that reaction. And I felt bad that my husband was feeling a bit uncomfortable, but I didn't feel bad enough to leave. No one was going to convince me not to see this doctor because I was hoping he would be the one to give me the answers. Maybe, as the head of gynecologic oncology at one of the top teaching hospitals in the country, he would have seen everything, and maybe, just maybe, he had seen something like this before. Everyone up until this point had given me nothing useful to make me feel that in the end I would be okay.

We were finally escorted to a consultation room where we met Dr. Schink and a resident. Dr. Schink asked me how he could help me, and I told him flat out, "You're going to be my doctor in case I need a hysterectomy. So you see me and I see you, and now I am your patient."

My blunt introduction didn't faze him. He asked me why I thought I might need a hysterectomy, and I told him about all of my fears. I told him I had been diagnosed with a complete placenta previa and was fearful of it becoming a placenta accreta, and then "I would need to have a hysterectomy and I am afraid I will hemorrhage to death."

"Okay," he replied matter-of-factly. "Have you been on the Internet?"

"Yes," I replied, "but this is what I believe is going to happen to me, and I can't get it out of my head."

"The likelihood of that happening is very rare," Dr. Schink said, "but I suggest you get an MRI at 35 weeks, and if we see the uterus and placenta permeating each other, I will schedule myself on the day of your C-section and we will take care of it then."

I left his office feeling much better that someone was

finally looking into my concerns, however rare they seemed to be. Dr. Schink was so calm about everything that it felt to me like he was in control and not at all concerned about what he had heard. I left his office feeling like I would be fine. But that feeling would be short-lived.

I called to talk about an MRI with Julie, who said that her protocol with her patients was to do one at 32 weeks, not 35 weeks, as Schink had suggested. I decided to follow her protocol because then I would know sooner than later. I did the MRI on May 2, and it came back negative for a placenta accreta. Jonathan said to me, "Don't you feel better?" "No," I said. The sense of doom was not going away. Instead of alleviating my fears, the MRI strangely heightened them. If it had shown a placenta accreta, at least my premonitions would have been tangible and we could have put a plan of action in place.

I couldn't kick this "crazy" sense of foreboding, and the list of people who would listen to me talk about it was diminishing. I asked Julie whether a placenta accreta could show up later, and she said she'd never seen it happen.

That night I tried to sleep, but the vision of organs melding together was still clouding my brain. Jonathan was frustrated: he had just seen the MRI results, and he insisted that tests don't lie. "Honey, you don't have an accreta. Please try to get some sleep. You need to rest. You can now cross that off your list of fears. It's not going to happen."

I desperately wanted to believe him. Intellectually, I knew that he was right, but deep down inside of me it didn't feel like it was true. I had no idea why. I was still scared.

I called my sister the next day. Michelle is 12 years my senior and had a hysterectomy in her thirties. I had not gone into too much detail about my premonitions with her because she would have divulged all the details to my parents. I didn't want the daily calls checking on me—it would have only added more stress for all of us. I told my sister the only information I knew she could handle. I wanted to convey a message that I was nervous, but not scared to death. I told her I was afraid I would need a hysterectomy.

"Why? Who told you that? Stephanie," she said in her motherly tone, "you will not need a hysterectomy." Then she ran through a litany of other possibilities. "Did you get your thyroid checked? Did you see a specialist about it? How is the baby? Is there something you are being told is happening? You need to take it easy, relax . . ." etc., etc., etc. It was a little bit of a relief to just hear her concern, but she didn't have all of the facts and I didn't want to go into them, so I just let her keep talking. "If you are to have a hysterectomy, even though you will not need one, make sure you keep your ovaries."

I laughed to myself. I understood what she was saying because she'd had a complete hysterectomy in which they removed her ovaries, and she had been on hormone replacement therapy for decades. It was terrible and painful to watch. I laughed because my ovaries were the last thing I was worried about. Michelle told me to stop being negative and to call her with any other questions.

The fear continued to build day after day. At one point I worked myself up into a full-on asthma attack and was convinced I was bleeding internally. Jonathan was in New York, and I called him in the middle of the night to say I needed to go to the hospital. Scared, he called his friend Alan to drive me as he searched for the earliest flight to Chicago.

Alan had just had a baby and was sound asleep, but he raced over and became my surrogate husband for the night. He has a very twisted sense of humor, and he and I always have great banter. Immediately after showing up, he helped me to the car and said, "You look like shit."

"Thank you, asshole," I said. "I am dying here. Get me to the hospital."

I said that very sarcastically, but then, as we drove to the hospital, he said, "What's wrong?" I told him in my shallow-breathed voice that I couldn't talk much, but I was sure my feet were swelling and my blood was pooling inside of me. He looked at my feet and saw nothing strange, but I was breaking out in cold sweats, so he kept handing me napkins to dry myself off. When he noticed that my face looked like it had been drained of blood, he knew something was wrong. He got me to the ER in seven minutes flat and convinced me that nothing bad was going to happen that night, even if he didn't believe it. He made me smile when he said, "Nothing's going to happen on my watch. Your husband would kill me."

That assurance made me a little calmer, and I felt, for a brief moment, that I was being protected. In the ER, I was monitored for a few hours and then told, to our surprise, that there was nothing wrong. I was released. Alan called Jonathan before Jonathan boarded the flight to tell him it had been a false alarm. In my mind, though, it wasn't a false alarm. It was another warning.

I called my rabbi and told him I was feeling things physically that weren't real and I was fearful that those feelings were foreshadowing what would come later. In a fatherly tone, he said, "You just need to think positively. Hashem [G-d] helps those who help themselves, but you need to be positive. Negativity will just eat away at you. So try to turn things around." The pep talk didn't help, but in his defense, he didn't have the history of what I was doing to help myself and I didn't have the energy to go into it again. Positive thinking couldn't change the way my "gut" was feeling.

At each OB/GYN appointment during April and May, I continued to tell my doctor of my detailed visions of dying while giving birth. Julie couldn't console me, but she suggested that I talk to an anesthesiologist. After all, an anesthesiologist would be the doctor keeping me alive during the operation. So I scheduled a phone consult.

On May 17, I called the obstetric anesthesiology department at Northwestern Memorial Hospital, and Dr. Grace Lim answered. We talked about the delivery plan, and she walked me step by step through what would happen from delivery to recovery. I already knew what she was telling me. What I really wanted her to spell out for me was the emergency plan in a worst-case scenario. I started to divulge all of my premonitions. I told her that I had a feeling my placenta previa would turn into an accreta, even though the tests were negative for it, that I would hemorrhage and need blood transfusions for my rare O-negative blood type, and that there wouldn't be enough blood to save me. I went on to tell her that I envisioned needing a hysterectomy and that I was afraid I was going to need general anesthesia. I added that I was sure Jacob would be fine, but that I would not make it through the delivery. I would die. It wasn't just a fear. I had seen it.

Once again, I was scared that a doctor wouldn't take me seriously. Later, Grace told me that my call caught her off guard. She had never had a conversation with a patient who was so clear and knowledgeable about what was going on and who was taking extra precautions by meeting with specialists just because she had a "feeling." At that point, unbeknownst to me, Grace flagged my file and wrote up a "plan B" that included extra blood, extra monitors, and a crash cart. Although I didn't know it at the time, "Grace from G-d"—as I've referred to her since—believed my premonitions. She was the only person who really heard me based on nothing but her own intuition. And it was her plan B that saved my life.

Chapter 4

As the days wore on, I learned to live with the desperation. Actually, I wasn't really living because I was constantly thinking about dying. I just tried to go through each day as normally as I could, but I also found myself trying to absorb every moment with my family.

I would sit and watch Adina sleep, remembering how precious she had looked when the nurse handed her to me the first time. I had been amazed at her tiny fingers and how the love Jonathan and I shared had created this amazing being. Then my heart would literally ache as tears ran down my face and I thought about leaving her. It was very painful. I felt so alone. It seemed no one could calm me and no one seemed to believe me. At least that's what it felt like. I was in the middle of the tracks, the train was headed straight toward me, and there was no one to pull me out of the way.

The doctors had scheduled my C-section for June 6, and it was almost like I could see my death certificate next to Jacob's birth certificate. A few weeks before I was to give birth, I turned to Facebook in a cry for help.

STEPHANIE ARNOLD (MAY 20, 2013, 9:40 P.M.): Any of my friends/family Blood Type O-? I need to have blood stored in two weeks—things are a bit hairy—I think I would rather have a transfusion with the blood I know than the blood I don't know . . .

MICHELLE (MAY 20, 2013, 9:48 P.M.): shoot. A+ Sending you love and light Mama.

SHARE (MAY 20, 2013, 10:06 P.M.): Oh goodness! I hope you and the baby are ok. Thinking of you :)

JENNIFER (MAY 20, 2013, 10:16 P.M.): A- what's going on???

MINDY (MAY 20, 2013, 10:22 P.M.): Crap. I'm A+. What's going on?

ISABEL (MAY 21, 2013, 7:34 A.M.): I may be O. Can it be sent to you? Let me know.

STEPHANIE ARNOLD (MAY 21, 2013, 9:34 A.M.): Need O negative only . . . I'm a universal donor but can only receive O- . . . Yeah I think it can be sent . . . I will check w hospital when I am there today

ISABEL (MAY 21, 2013, 9:39 A.M.): I think I am O+

CAROLINA (MAY 21, 2013, 10:33 A.M.): I think my mom might be or my dad actually

DEIDRA (MAY 21, 2013, 7:06 P.M.): Joe is O. I will check whether he is - or +. Will be praying Steph.

Again, there was no medical evidence that I would need extra blood. The tests all said things were normal. But it felt like what the visions foretold was already happening in my body. I was feeling pain where doctors said there was no evidence of it. I had sweats although tests proved I wasn't running a fever. I felt blood pouring out over my organs, but ultrasounds showed nothing. People were seriously worried I was falling off the deep end, but I never relented. I can't explain why, but I knew these visions were not a figment of my imagination. They were foreshadowing what would happen in a couple of weeks, and I knew it was coming. I just didn't know how to stop it.

I was getting nowhere with the testing, and being constantly dismissed by doctors was changing my emotions from sad to incredibly angry. I was angry with my doctors that they couldn't find out what was wrong. I needed them to fix it, so I could stop spinning my wheels. I was frustrated that I could find nothing in my research to prove that what I was feeling was real. I needed proof, but was I really going to have to die in order to prove it? I was angry that I was wasting so much energy around this when I should have been enjoying my time with my family. And now it was too late. I wanted more time. I was looking forward to our future, but the future was here, now, and it was about to be over.

Because I believed with such certainty that I was going to die, I began to write good-bye letters to friends and family.

Dear Jonathan, You have made me the happiest woman in the world . . .

Dear Adina, Please know that Mommy will always be around you . . .

Dear Valentina, When your dad brought you into my life . . .

I stopped some of these letters midway because I felt that, if I left them unfinished, I would survive. Logical or not, that was how I dealt with the fear.

Interestingly, there was one letter I managed to complete. On April 28, I sent a letter to the embryologist who had helped create both of our children. I wanted her to know what an important role she had played in my life and how I missed our close friendship. We had lost touch over the past year, and I wanted her to know how great an impact she'd had on our lives.

. . . I wanted to write to you about how I felt in case I don't make it through delivery if I do have to have a hysterectomy with all of the risks to an accreta. I am writing my letters to Jonathan, to Adina, to Valentina, and to our unborn son & to a few friends and family members as well. You are included in this because you are important to me and if I don't make it, you know you were in my thoughts these last few weeks and the last year for that matter.

I started many times to try to write my father, but over and over again I couldn't get past "Dear Papi." I was stuck. I had so many things I wanted him to know. How he was the first love of my life. How he taught me to be a giving and caring person. How he instilled in me kindness and self-respect. I didn't want to believe what was about to happen to me, and I most certainly didn't want to leave him. When I thought about how he and my mom would feel when I died, my only relief was in thinking that my children would be their connection to me.

My father, Ralph, was always my go-to guy and my rock. He was there for me financially, emotionally, and spiritually. My dad is Cuban, and spirituality is a big part of that culture. His mother was very spiritual. My Grandma Ida was considered a *bruja* (witch), but in a good way. She would see things that would later come true. One family story that circulated throughout the years was that my grandma once was very ill and lay down for a nap. When she awoke, she told her husband, Abe, that her late mother had visited her in a dream and given her some medicine that she felt going down her throat. She was miraculously better immediately.

My dad, a judge and a lawyer, was very analytical and pragmatic. He would tell me that what happened in his mother's crazy stories might just be coincidence, or imagined. He said he didn't believe in that stuff. Yet, when he thought he was all alone, I would catch him asking for his late mother's help. He even talked to his brother-in-law, my Uncle Marvin, who was also deceased. My dad would ask them to watch over his family, give him guidance, and keep everyone safe from harm. I remember asking him why he prayed to them and not to G-d, and I will never forget what he said: "I don't know if I can count on G-d, but I can always count on Grandma Ida and Uncle Marvin."

Interesting words. I had a special connection to these

family members as well. When I was 12 years old, my family was on a trip to Las Vegas to celebrate the New Year. Grandma Ida was back in Florida. Just after 10:00 P.M., a sharp pain hit my heart. All I could think about was my grandmother. An hour later the phone rang. Grandma had just died of a heart attack.

When I was 20 years old, my Uncle Marvin had gone with us to temple for Rosh Hashanah. Marvin and I were extremely close, probably because we were both nonconformists. I loved him so much, but interestingly, I had never told him. Marvin was Jewish, and he had married a woman who was not. His children were raised outside of the Jewish faith, and that always bothered him. That day I remember him telling me that he was sad that his children weren't Jewish because he wouldn't have anyone to say Kaddish (the prayer that mourners say for the dead) for him when he died, a weird thing to say when you are a healthy man who runs five miles a day.

I got a strange, overwhelming feeling when we hugged good-bye in the parking lot that day. And for the first time I looked him in the eye and said, "Uncle Marvin, I love you, and I promise I will say Kaddish for you whenever you need it." He smiled and thanked me, and we parted ways, saying we would see each other for Yom Kippur services in 10 days. But I knew that was the last time I would see him. He died two days later, and I kept my promise.

In the weeks leading up to Jacob's birth, I kept writing those letters. Now, on May 30, I was driving to the hospital to probably give birth a week before my scheduled C-section and wondering who was going to say Kaddish for me.

I called my father in the most tranquil voice I could muster to tell him I was going into labor. My father had a weak heart, and I knew I needed to speak calmly to him so he wouldn't panic. He said he would let my sister and the rest of the family know, and he asked me to call him after Jacob was delivered to let him know everything was okay. He then added, "Good luck, sweetheart, we are excited to meet our grandson."

Tears filled my eyes. I said, "Good-bye, Papi, I love you." Five words filled with intense love and now anguish.

At the hospital labor and delivery entrance, I was admitted quickly. A nurse hooked me up to the monitoring equipment and told me to relax. Right. Adina was with me in the room. So was Tessie. My friend Jodi met me at the hospital and told me she would be there as long as I needed her. She always makes me laugh, and I needed her stupid sense of humor to take my mind off of things. At one point I walked to the bathroom and passed a clot on the floor. Jodi, with her quick wit, said, "I think you dropped something." She made me laugh so hard. Tessie assured me she wouldn't leave Adina's side.

At that same moment, hundreds of miles away in New York, Jonathan was at the New York State Bar Association antitrust meeting. The host had just called the meeting to order when Jonathan's phone started to vibrate. He walked out to the hallway, answered, and heard my panicked voice. "I am on my way to the hospital. We are going to have this baby today. Get here as fast as you can." After expressing how much we loved one another, totally, completely, and forever, we hung up.

Jonathan hopped into a cab, made a flight reservation en route, and stepped on board just as the door closed. He would remember that trip vividly as he recounted it months later.

"The moment we hit 10,000 feet, I logged onto the Internet and started Skype-chatting with Steph," he said later. "I felt horrible that I was not with her and she was scared and alone. She was updating me constantly on her condition. She told me she was being watched closely, and while I was upset that I wasn't there to hold my wife's hand or console her, I didn't think we faced danger. Steph, on the other hand, felt great danger. She kept reiterating her fears and begging me to get to the hospital without delay. Steph was happy that she was able to persuade the doctors to delay the C-section until 1:30 P.M., so I would have time to get there."

Then there was a reality check. Julie changed the plan. Everything was quiet in the operating room, so she thought it was best to start the procedure even though Jonathan was still en route. I continued typing as frantically as I could, like a high school student racing against a clock while taking a timed test. Where could I start? How should I start? What could I say knowing these were the last communications I would ever have with the most important person in my life?

I wrote, "You are the best husband and the greatest father and your children will always know it." I continued: "I love you with all of my heart. And no matter what happens, you have made me the happiest woman in the world." Still focused on the logistical questions, he asked where he should meet me, and I said, "The 8TH floor, Recovery . . . hopefully." That was it. Our conversation was over. I had to shut down my computer before I could see his reply.

"Somewhere over Ohio, the plan suddenly changed," Jonathan said later. "Steph IM'd me that they were prepping her for the C-section. Why the sudden change? All she did was repeat over and over how much she loved me, that she thought our marriage was wonderful and . . . nothing. The line went dead. The IM connection went dead. My heart sank."

My fear was palpable now. I looked at Adina, wanting her to remember me as a happy woman who was desperately in love with her. I kissed her a few times. My friends assured her I would be back, and my doctor gave her a hug and said, "We will be back with your brother soon." As I was being wheeled out of the room, looking back at my daughter's face, I held back the tears as I realized this was the last time I was going to see her.

Chapter 5

THE FEAR WAS EXCRUCIATING. I desperately needed to hold Jonathan's hand, to hear his voice. I knew that if he was there, he wouldn't let anything happen to me. As I was being wheeled down the corridor by my doctor, I tried again. "Julie, there is something wrong with me, put me under general anesthesia, the baby is fine, but I need to be under general anesthesia." She told me to calm down and reminded me that all the tests were negative for any problems. "I'm sure you are just nervous because Jonathan isn't here, but it will be fine. I will take care of you."

D-day was here, and under the bright lights of the operating room, surrounded by strangers who would hold my life in their hands, I finally let go. I put myself in my doctors' and G-d's hands. I was being wheeled into the room where my son's life would start and mine would end. I was sure of it. Attached to IVs and strapped to a gurney, I felt like I was being held hostage and there was no way out.

I remember a lot of people in the room, preparing for the surgery, and I also remember someone putting soap on my belly where the place for the incision would be marked. They had to wait three minutes exactly for the soap to dry before they could begin the procedure, and during the entire 180 seconds everyone was silent. Everyone but me. It was so tense that I tried to make light of the situation by asking everyone, anyone, to tell me a joke. They weren't laughing. They were ready to begin.

And that's where my story ends. Or ended. Or could have ended. I gave birth to a beautiful baby boy named Jacob, and seconds later I died.

"I WAS GETTING ANXIOUS to see Steph," Jonathan recalled. "The plane couldn't get there fast enough. I couldn't wait to meet Jacob and kiss my wife. As soon as the plane landed, I texted Julie, asking her how everything went. She answered, 'Jacob is fine. Stephanie is . . . stable.' That wasn't what I expected.

"I arrived at the doors to the delivery room just as Julie was walking out. She had her scrubs on, covered in blood. My wife's blood. I went into 'emergency action' mode. This was my natural inclination, and my Air Force pilot training reinforced it. Whatever emotions I possessed were tabled and replaced with cold analytical thinking and evaluation.

"Before I could ask a question, Julie told me that Steph suffered an extremely rare pregnancy complication and that she was in recovery and stable. She said it was an amniotic fluid embolism, or AFE. Extremely rare. So rare the doctors, who were all very seasoned, had never seen one in person, only read about them in medical books. Julie handed me a bag of Steph's jewelry, including her wedding bands, and told me to wait for the anesthesiologist. They ushered me into a conference room and Dr. Nicole Higgins arrived.

"Nicole explained it's a condition where material from the baby gets into the mother's bloodstream. It could be a drop of amniotic fluid, a hair, or even as small as a cell. It somehow enters the mother's bloodstream, and if you happen to be allergic to it, it sets off a chain reaction, breaking down the structure of the blood and in turn causing cardiac arrest, shutting down the organs, and destroying everything in its path.

"I started peppering her with questions. 'What is Steph's current state? What are the current threats she faces in the next hour? Two hours? Twelve and twenty-four hours?

What treatment is she receiving? What judgments are being applied? What is the alternative?' Nicole exhibited no emotion when answering the questions. This was her comfort zone too. We communicated just fine. Later, she told me how much she appreciated the absence of emotional drama with me as well as the cold analytical assessment we undertook.

"Nicole continued. Amniotic fluid embolisms have two phases. The first is that the body goes into anaphylactic shock and it flatlines. Some studies show as many as 50 percent of AFE victims don't make it past that point. But if you happen to be lucky and get to phase two, you could bleed out because the clotting capability of your blood stops working.

"She said that both Jacob and the placenta were delivered perfectly, but after that it all went downhill. Steph had a seizure and flatlined. She was clinically dead for 37 seconds before they resuscitated her. Nicole explained that her platelets went from 236,000 to 12,000 in less than 12 minutes. Steph's blood could no longer clot, and she was bleeding out. She would be given 60 units of blood and blood product (platelets, red blood cells, etc.) over the next several hours."

Anesthesiologists will tell you that their worst nightmare is something called DIC, which stands for disseminated intravascular coagulation—the inability to clot. If your blood cannot clot, you will hemorrhage to death. Doctors sometimes refer to it as "Death Is Coming." I was on the edge.

They wouldn't know for days if I would survive and if I did, whether there would be any neurological damage or other consequences of the initial attack. All Jonathan could do was educate himself on what had happened in order to make the best health decisions on my behalf, and then wait.

To start trying to make sense of what he was hearing, Jonathan got online and Googled "AFE" to find out more about it. He read that it's completely unpredictable, unpreventable, and in most cases fatal. He saw that it was rare. Only 1 in 40,000 births. There weren't many stories of survival. Almost everything he found talked about support for widowers. He kept thinking it just wasn't possible. He was not going to lose me.

"As I waited for word on Steph's condition, I knew I needed to see Jacob. It felt very lonely taking the elevator up to the nursery. My wife was in dire straits. I had never even seen Jacob, much less held him, and I was going to see him, alone.

"Jacob was a beautiful sight to behold. When I walked up to the secure door to the nursery, I saw the nurse putting him into his cradle. She had just fed him, and he was already asleep again. She buzzed me in. While I sterilized my hands, she told me what a wonderful little boy Jacob was and wheeled him into an adjacent room for us to have some private time.

"As soon as I picked him up, the suppressed emotion came flooding out. I rocked him back and forth and wept.

I assured him that I was going to make sure his mommy would be fine, all while not knowing what could possibly be done to alter the trajectory my wife was on. In retrospect, I was lying to him and, more importantly, to myself.

"I spent over two hours with him, holding him close. I took off my shirt so he could have skin-to-skin contact. I knew that to be important. And it should really be mother and child. I would have to do, under the circumstances. We bonded immediately, and he looked up to me with his beautiful little eyes and snuggled up into my neck and shoulder. He slept easily and soundly, and it felt perfect. Almost.

"That feeling of perfection gave way to my desperate concern for Steph. The docs told me to meet them over in the surgical intensive care unit in a few hours. I left the nursery and started the 15-minute trek to the ICU.

"It was the first of many trips that would take me down the elevator, across the pedestrian bridge, through the parking structure, over a second pedestrian bridge, and up another elevator to the ICU. I would come to like the walk because it gave me time to think and review all the dozens of new facts flowing into my head each hour. It also gave me time to steady myself for the terrifying juxtaposition of the promise of new life, with all the wonders it brings, against the threat of death or permanent paralysis, degraded neurological function, and other side effects of the horrible AFE that struck my wife.

"I arrived at the ICU at 3:15. Steph was not there. The

chief resident told me that she had not even left the OR to start the move to the surgical ICU. I should come back in no less than an hour. I was back in 30 minutes. I was in no state of mind to spend more time elsewhere. The details of an AFE were making my mind spin. I needed to see my wife, and there was no way to do so.

"I went and got some food. I wandered around the hospital aimlessly. I didn't call anyone because there was nothing good to say and no reassurances I could give. All I would accomplish was scaring people needlessly. I sucked up the calamitous situation and kept it inside. She was dying on the operating table, and I was dying inside.

"I returned to the ICU at 3:45. No Steph. Back again at 4:00 P.M. No Steph. Finally, I planted myself in the room assigned to her and waited. At 4:35 P.M., Steph was wheeled in through the back entrance to the ICU. She looked horrible. Scary. My heart sank as I scanned her from head to toe. She was pale, with all the color drained from her face. One doctor was using a device to pump air into her mouth to keep her oxygenated. She was hooked up to more than a dozen drips. Her wrists and ankles were strapped to the rails of the gurney, presumably to prevent her from shrugging off her IVs. Blood seeped through her gown and bandages.

"Blood was being infused into her body. They had put my wife into a medically induced coma."

Jonathan later said that when he saw me, it almost

brought him to his knees. Apparently, I was unrecognizable. Not just because of the bloody blankets, but because my body was unbelievably puffed up by all the fluids being pumped into me.

"As they started preparing to transfer Steph from the transport to her ICU bed," Jonathan said, "the [ICU] fellow, who seemed to be the person most in charge, started to manhandle Steph's limp body. When he realized I was watching, he asked me to leave and come back in 45 minutes. It felt wrong to leave her, but the combination of his request and seeing my wife in pain during the transfer resulted in me walking to the waiting room for a few minutes. My mind was racing through all the permutations of what happened, what needed to be done, and what outcomes lay in my future."

Jonathan told me later that he was on autopilot, trying to figure out how to save me. He went from doctor to doctor, telling each one that if I needed a hysterectomy, Dr. Schink was our doctor. At that time I was stable, and the doctors told Jonathan that a hysterectomy was unlikely, but they took note of the referral and went back to check on me.

I think he was worried he was being overly cautious, but he was happy and relieved that he spoke up when the doctors told him they thought it wasn't going to be necessary. One less complication to worry about and at least one premonition he was happy wouldn't come true. Or so he thought.

"When I finally was allowed back into the ICU," Jonathan said, "Steph was in the final stages of being transferred to her new bed. The doctors and nurses were checking all the hookups to Steph, and there were many. She had four poles, and each pole had three to four bags of drugs hanging off. The bags were slowly dripping into her. Some were antibiotics, others were blood pressure meds. An automated blood pressure cuff took her blood pressure every few minutes, with the most recent reading announcing itself on an overhead monitor. Her heart rate was updating even more frequently. In addition, she was hooked up to a breathing apparatus because she was in a coma and could not breathe by herself. And to top things off, a large bag of blood was hanging from the ceiling, slowly infusing into her. Outside the room, several spare bags of blood were sitting on a machine that was slowly rotating, presumably to keep the blood moving so it would not go bad.

"It was an ugly sight. Every few minutes Steph would make some noise and move her arms and legs a bit, only to be constrained by her wrist and ankle restraints. The restraints probably made it feel like prison, but it was soon obvious that Steph would pull herself off of her many IVs in their absence.

"I sat down in my chair and watched for many minutes, taking it all in. There was an immense amount of information flowing onto the screens, and it was updating rapidly. Doctors and nurses needed to be able to digest all this information in a flash and then make the correct decision on what to do or what to change, or to not do anything at all. My wife's life depended on it. It reminded me a bit of my days as an Air Force pilot flying a military plane and getting an in-flight emergency. When such an event occurred, the instrument panel would light up with multiple systems failures, followed by the plane losing power, altitude, speed, and more. All in a fraction of a second. We trained on what to do in these eventualities. When they actually occurred, we had to execute immediately, reflexively.

"The doctors and nurses were executing. No doubt about it. Steph was under constant attention on the afternoon of Thursday, May 30, 2013. A dedicated ICU nurse, the chief resident and ICU fellow, along with additional staff, were devoting the vast majority of their time and attention to Steph in her first few hours in the ICU.

"During that time I received many briefings and updates from various medical personnel. The news was guarded. Everyone emphasized that her incident was extremely traumatic. No one said, 'Lucky to be alive,' but that was quite evident. They said that Steph had a long way to go to get out of danger. No one said that she was at risk of permanent neurological deficit or at risk of other adverse side effects that would last a lifetime. Everyone mentioned that Steph's heart had stopped for less than a minute, which they said was good, and that her blood remained oxygenated for the duration. I interpreted that to mean Steph had a chance, but only a chance, of coming out of this okay. Over the subsequent days the doctors would be more explicit about the risks, but I was already prepared.

"My voice mail was piling up. Family members wanted updates. I called Steph's parents and told them Steph gave birth and that Jacob was fine, in fact, doing great. And I told them Steph was sleeping. All true, of course, with one great omission. But I needed to focus on Steph and not try to manage family issues. I had to make a decision, as hard as it was.

"Shortly after that call, Steph's sister, Michelle, called me to say that she was making arrangements to fly up to Chicago that night. It was less of a question and more of a statement. I gave her the same update that I'd given her parents.

"One piece of good news was that Steph was looking better. The hemorrhage seemed to be slowing down. She had already received a huge quantity of blood transfusions during birth. A port, protruding from her stomach, was draining all of the blood that was pouring into it. At around 4:30 P.M., the bag was nearly full and was swapped out for a new one. That new one was filling up very slowly. The doctors were happy about this. It meant that the internal bleeding was getting under control.

"Things in Steph's ICU room had settled down to a steady level of frenzy. In fact, it felt like a low level of frenzy, but that was probably a consequence of the frenetic attention she received in her first two hours there.

"I had the fortune or the misfortune of WiFi in the room, so I hopped on the Internet to find out all the information I could on AFEs. With a few clicks, and after 15 minutes of research, I got myself pretty scared. AFEs are a leading cause of maternal death around the world. Most women do not survive a full-blown AFE. For those women who do survive, the health outcome is generally poor, permanent paralysis or permanent neurological deficit being the likely result.

"The AFE Foundation website, afesupport.org, hit my radar in short order. I was immediately drawn to one of its main tabs: 'Grieving Fathers.' There I read story after story about husbands losing their wives and even their newborns. There were many resources directing you to widower support and parental loss support. I felt the horror overcoming me.

"Was this going to be my future? My mind was racing but was quickly interrupted by the stepped-up activity in the room. Steph's second bag of blood was running dry, and a third was being hooked up. Almost no blood had drained out of her abdominal line. Good, or so I thought.

"Her belly was getting harder and harder to the touch. Steph was still hemorrhaging. Her body was positioned in such a way that the blood in her stomach wasn't draining the way it should, which was why we all thought things were getting better. But they weren't. The amount of hemorrhaging was unacceptable. She needed a hysterectomy, and she needed it right then. They paged Dr. Schink."

The pathology report on the uterus validated my earlier premonition—a placenta accreta *had* formed. My husband was in disbelief. This was way too much for him.

Dr. Schink performed the hysterectomy, and the surgical team packed me with gauze. The bleed was finally under control, but the ordeal was far from over.

Chapter 6

As the hours went by, my body clung to life with the help of machines. Jonathan sat in the room wondering if I would ever wake up from the coma.

"Usually between 9:00 P.M. and 5:00 A.M. the ICU was extremely quiet," he remembered. "It was at these times that I allowed my mind to drift from its work of assessment, analysis, evaluation, etc. My mind instead focused on all the unknowns: Would my wife survive? If she did, would she have permanent impairments? If so, what would they be? What would be required of me? How on earth could I raise our children without her? Those were the scariest times for me. And I came to dread them."

The doctors would come in and tell Jonathan that they didn't know how well I would recover. He feared I'd have brain damage. He saw how swollen I was becoming due to kidney failure and didn't know whether I would stay that way. I had ballooned to three times my size. By then, he was well aware of the recovery statistics of an AFE. They were not good.

Jonathan tried to distract himself by doing calculations on the whiteboard inside my ICU room. He was calculating the "frequency distribution" of AFEs in any given year at this hospital, using formulas to figure out the probabilities of this rare occurrence. Not that it mattered, because we were smack in the middle of this rare occurrence, but it helped him stay busy. Doctors would come in and out, look at his math, and shake their heads, trying to figure out what he was doing. They would check my vitals and tell him once again that they didn't know how I would fare. None of it was helping Jonathan, so he went into crisis management mode, determined to believe I would be fine. I had to be fine.

In the meantime the calls kept coming. Jonathan continued to say, "The baby is great, Steph is sleeping right now, she is unavailable to speak." He didn't know what else to say, but he knew he didn't want to tell everyone how dire the situation was until he fully knew what was happening. Friends and family would persist and call back every couple of hours. My father called several times that day, and Jonathan would say, "She is in with the doctor right now, but we will speak to you later." He never lied, really, but he didn't exactly tell the whole story. He didn't want to scare my father, and he didn't want to field calls every hour. He needed to concentrate on me without any distractions.

My sister Michelle arrived at 2:00 A.M. on Friday, May 31, thank goodness. She and my brother-in-law Roy didn't know what had happened, but she had a feeling that something wasn't right and hopped on the earliest flight that would get her from Florida to Chicago. Jonathan knew he needed to tell them what was going on, so he met up with them for breakfast across from the hospital around 7:30 A.M.

My sister immediately noticed how terrible he looked, and Roy joked about him never getting another good night's sleep now that we had given birth to a boy. Jonathan wasn't in the mood to smile. He sat down and said, "Your sister's condition is very serious. I was hoping to put this off another day or so until I knew more . . . but here you are. She is in the ICU and had an AFE."

He explained to Michelle and Roy what that meant. He reassured them that Jacob was fine. After breakfast, as they walked over to the hospital to see me, he tried to prepare them. "It is a really horrific sight to see her in the ICU. And no matter how bad you think this is going to look, it is going to be worse—much worse—when you see it."

Michelle said later that when she first saw me she was confused. She thought she had walked into the wrong room. "I was staring at a 300-pound person, and that wasn't my sister." I looked nothing like myself. As reality settled in, she too went into crisis mode. There were no tears, and as she will readily tell you, there were no fears. There would be only one outcome in her mind. I had to survive.

After getting briefed on my condition, Jonathan took Michelle to see Jacob and to hold him. She said that was the hardest part. "It was really sad to see Jonathan holding Jacob without Stephanie. It was an incredibly sad time when we were supposed to be happy. A complete dichotomy."

Jonathan made Michelle swear to not tell our parents until they knew more. I get it now, and so does my sister. It was the right thing to do. But in the moment my sister felt strongly that if I was going to die, my parents should have the chance to see me one more time to say good-bye.

My sister is very good under extreme pressure. She shuttled back and forth between the maternity ward and the ICU. She asked Jonathan where the "stuff" was for Jacob and me. "What are you talking about?" he asked, perplexed. Michelle said, "His clothes, blanket, plush toys, a bathrobe and toiletries for Steph, their *stuff.*"

Jonathan hadn't been thinking about any of that and told my sister he didn't know. She knew how organized I was and couldn't understand how I could have come to the hospital without at least some personal items for myself. Little did she know that I hadn't felt as though I'd be needing them. I hadn't thought I was ever going to leave the hospital. In her typical way, Michelle went out and bought some items for me and everything Jacob would need for the first six months of his life. Including shoes.

She held Jacob any chance she had. She made sure the nursing staff was well fed and attended to because they were the ones keeping Jacob alive and very happy. She changed him into street clothes and wrapped him in his own swaddle blankets so he didn't look like an abandoned hospital baby. I think she just wanted him to feel loved by his family and to know that he had a mother, if only through her sister for a while. It still makes me cry to this day when I think about those moments.

As hours went by, and soon days, I lay still in the ICU, surrounded by machines and attached to tubes, with my family sitting vigil. My parents would call every four to five hours, and my sister and Jonathan were running out of excuses. Michelle called our cousin Sari to talk it through. She explained that she felt caught in the middle between respect for Jonathan's wishes and a need to tell her parents so they could fly to Chicago. Sari and Michelle devised a plan that would keep them busy until it was the right time to divulge the truth.

They knew that Mom would know what to do. She always knew what to do in a very heavy situation. And our mom would be able to be strong for everyone. Michelle needed for Mom to know. She never went against Jonathan's wishes, though, and she didn't even call our mom. After three days of not being able to speak to me, my mother's intuition was running in high gear. She called my sister and said, "Michelle, what is going on? Do I need to come up there?" Michelle mustered up all the strength she could to avoid blurting out what had happened and just said, "Yes, Mom, I think you need to come up now. Yes, now." And with that, my mom was en route.

My mother, Charlene, is a pretty strong woman. At 75, she exercises for six hours every day. She is the daughter of a Jewish gangster who used to work for Meyer Lansky, although she constantly corrects me and tells me her father was a "legitimate businessman." She is tough as nails, and although at first sight you can't help thinking she is a tiny woman who would blow over if a strong wind headed her way, she is the complete opposite. She has inner strength and gumption that is unparalleled in anyone I have ever known. Nothing stops her or intimidates her into not doing what she wants to do. Nothing.

Years ago, when I was 13 years old, I was in Vegas with her when she sat down at a private poker table to play with some very substantial "whales." They were smoking their cigars and blowing the smoke out when she sat down at their all-men's game. They started cursing and saying this was not a place for a little lady. I was intimidated, but my mother could have cared less. No one was going to tell her she was out of her league.

My mother had only a handful of chips, and these guys had huge stacks in front of them. She even had the courage to look at one of the guys and say, "Do you mind putting your cigar out or blowing it the other way?" I can tell you, he was not pleased to hear that. I said, "Mom, you okay?"

"Yes, honey," she said. "Go see your grandfather and meet me back here in two hours." I was nervous for her, but I left.

Wouldn't you know, when I went back all of those whales had barely any chips in front of them, and guess who was loaded up? That's right, my frail, tiny mommy. She said, "My daughter is back, I have to go. See ya, fellas, have a good night." The guys looked at her and said, "C'mon, Charlene, we are just getting warmed up, stay and play," but my mother was shrewd. She was an excellent poker player, not just because she knew instinctively when to bluff and when to fold, but because she knew exactly when to stop playing. Her "momma didn't raise no fool."

I guess I learned that tenacity and intuitiveness from her. Now, in the ICU, I wasn't bluffing, and I wasn't going to stop fighting. This wasn't a game. This was real, and I was playing for the highest stakes of my life.

My mother had worked as a volunteer in hospitals, and

she had seen some pretty sick people. Nevertheless, Michelle, Roy, and Jonathan had to prepare her for what she was about to see. Michelle met my mom in the hospital lobby and met everyone outside of my ICU room. They told my mom that her daughter was inside that body, but the body didn't look like mine. They said that she would be seeing a lot of machines and hearing many sounds, but that I would be quiet. She would see blood everywhere, on my neck, on my arms, on my legs, through drainage tubes in my abdomen, but I would not be feeling any pain.

My mother, getting impatient, said, "I have seen worse things. Let me go in and see my daughter." Before they went in, Jonathan asked her not to tell my dad until they knew my prognosis. She agreed and proceeded to walk into my room.

Jonathan and Roy walked in with her, one on either side. When she caught a glimpse of me in the bed, she almost passed out. They grabbed her under the arms and stood her up. In that moment, she told me later, she was brought back to the time I was five years old and had a terrible case of sinusitis. The doctors didn't know what it was at the time, so I was hospitalized. Half of my face was swollen and abnormally drooping down. Uncle Marvin came to visit me and asked my mom, "Where is Stephanie?" She said, "Right there in the bed." He said, "That's not her, it looks nothing close to her." And that is exactly how my mother felt now, standing in front of me 36 years later. The body in the bed looked nothing like me. She said the color of my hair was all she recognized. She was going to be sick.

She immediately phoned my father crying and couldn't hold back. The first thing she said was, "We lost her." As hard as everyone had been trying to keep the bad news away from my dad, with his heart condition, she couldn't contain herself. Luckily, she followed that up with, "She's in the ICU and in a coma," or my dad would have lost it himself.

Jonathan and Roy told her to sit down and got her some water. After a few moments, she realized she had to be strong. She walked over to me, held my hand, kissed me on the forehead, and told me stories about everyone in the family.

All my life she had been wishing I would have a kid just like the tough, strong-willed, and stubborn kid I once was. But in that moment she would have given anything to hear me fight with her. She wanted the doctors and Jonathan, Michelle, and Roy to know that I could hear, that I could talk, and that I was feeling every touch.

Meanwhile, down in Florida, my cousin Sari was telling my dad that he needed "to drum up as much courage as you can find and go to Chicago." My father hated hospitals, but more than that, he knew he would hate to see his baby at death's door. He knew he needed to go, but was too scared to see me. After a few days, though, he agreed to fly up.

When he arrived, he walked into my room, bent down and kissed me, and said, "Papi's here, everything is going to be okay. Stay positive and stay strong." Then he swiftly left and went to the waiting room. The sight of me had been too much for him. He needed to pray. "I asked G-d," he said later, "if there was a chance that he was going to take you, to please take me instead." Then, out of breath, he started to shake and needed his heart medication. He went to get a cup of water, then sat back down to recover.

My brother Mark had come with him. Mark is an Orthodox Jew, and he made sure to bring a community of prayers with him to ensure my recovery.

My room in the ICU had become crowded with love. My sister put pictures of our entire family up on the walls so that, if I woke up and no one was there, I wouldn't feel alone. My mother would sit beside me, holding my hand, then leave to go relieve her stress by doing yoga in the waiting room. Mark would pray, reading through the Bible front to back, and around my bed he would pin "evil eye" charms, which are supposed to protect you. My dad's anguish was starting to affect his heart in more ways than one—he was having trouble breathing—so he would stay for only short stints. While he was in the room, though, he would pray for help from Uncle Marvin, Grandma Ida, and G-d.

And in the building across the pedestrian bridges, the love of my life held our precious baby boy and told him he couldn't wait for him to meet his mom. It was a promise Jonathan wasn't sure he could keep.

Chapter 7

MY ROOM HAD BECOME a makeshift shul. My family prayed. Jonathan prayed. Hospital staff and my doctors stopped by to pray. People of all different faiths prayed. They were all coming together, sitting vigil, hoping and praying for a miracle.

Jonathan got used to the rhythmic low hum of the machines breathing for me, peppered by the beeps from the blood pressure monitor as my numbers sometimes soared for a moment, then dropped back to normal. He was constantly reminded that I was not able to do any "living" unassisted.

He would think about all our plans that might never come to fruition.

This might sound morbid, or maybe just practical, but before all of this happened we had dreamed of being buried next to each other. We wanted to find the right cemetery for us and the perfect plots where we would be together for all eternity. Would he now have to make that decision on his own? He was also thinking about his children: How would he handle their daily routine without a mother of the house? Would he be asking for help from everyone he knew? He had to shake off these thoughts and decide that it was never going to happen. I would be fine. I had to be fine.

Jonathan had to face a lot of people and deal with reality sooner than later. He had to explain to the New York Attorney General's Office that he wouldn't be coming back to work as quickly as he'd thought and needed to take a medical leave of absence from his position. He had to prepare a mass e-mail to our friends and family because, by this time, the word was out that something must have happened since I hadn't returned calls from anyone for days.

Jonathan knew he was going to have to deal with the world eventually, but for now, he would keep the world at bay. He had to worry not only about me in one hospital but about Jacob across the street in another hospital. And he had to think about the bris. Jacob would soon be eight days old, and according to the laws of Judaism, he had to

have his bris, or ceremonial circumcision, on the eighth day.
This pact between a male baby and G-d is prescribed in the
Torah.

Our family rabbi told Jonathan that, as long as the baby
was healthy, the ceremony needed to proceed, even without me. The search began for a mohel—someone who is
trained to do Jewish ritual circumcisions—so Michelle and
Roy called their rabbi in Bogota, Colombia, who called the
Chabad community in Chicago and got the name of one.

Then my sister spoke with the operations people about
having the circumcision at the hospital so that even if I
wasn't there in mind, I could be there in body. They told her
it was rare to have someone who is not licensed to practice in
the hospital perform any kind of procedure there. And even
though a circumcision was a minor surgery, it was still a surgery. Jonathan felt there was no way it could happen. He is
a man of rules and order, and he knew this rule was in place
for safety reasons, so he wasn't going to challenge it. But he
didn't seem to know my sister very well, or the kind-hearted
folks at Prentice and Northwestern. They all bent over backward to make it happen.

And wouldn't you know it? Through Chabad, via South
America to the States, Michelle and Roy found the same
mohel I had made arrangements with weeks before when I
was planning Jacob's bris. Other rabbis came too. They told
Jonathan to pray. Pray for a miracle. Pray that I would be

okay. Pray that I would be there for our son's circumcision. And so he prayed.

The miracle came the day before the bris.

I woke up. Maybe it wasn't a true miracle, but the timing sure made it seem like it was. My vitals were getting stronger, so the doctors had started decreasing the medications that were keeping me in the coma.

The first thing I remember saying, as I looked down at my swollen belly, was, "Am I still fucking pregnant?" No one knew how much neurological damage had been caused by the AFE, but my husband said he knew at that point I was going to be okay because I knew why I was in the hospital and I was cursing.

He said, "No, sweetheart, you gave birth six days ago." That blew my mind. "What do you mean?" I was devastated. Where had I been? Tears were rolling down my face. He tried his best to explain, but I honestly have no recollection of what he said. Apparently I was still heavily medicated.

I cried and cried and cried. I cried about the loss of myself, and I cried because of the pain. The pain was immeasurable. My kidneys had started to fail, which had led to more tubes, more tests, and more pain.

I was in a daze as they wheeled me from one hospital room to a different room next door. They gave me the deluxe primo suite they gave to their VIPs. Room 1368. It was a corner suite with a view of Lake Michigan in one direction and

Chicago's famous Water Tower in the other. It was bright and cheerful, and as a "special" high-risk case on the floor, I would be calling it home for the next few weeks.

At one point a patient transporter came to take me from one appointment to the next on a gurney, and when he brought me back he said, "What did you have to do to get this room?" I said, "I had to die. So not worth it." I wasn't in the mood to explain, and I guess he got the message: he slowly backed out of the room.

Painkillers kept me from fully comprehending anything. I understood, kind of, what had happened, but I was told to not read anything about it just yet. I've always been a researcher and full of curiosity, but this time I wasn't arguing. I didn't want to know too many details. If it was anything close to what I had been visualizing for months, I didn't need to have that confirmed by reading other devastating stories about it. I needed my strength to heal.

Although it was hitting me periodically that I had thought this was going to happen, I still couldn't wrap my head around it. And the fact that I came out of it on the other side . . . well, I needed time to decompress. Unfortunately, I wasn't going to get much time to do that.

It had been seven days. Unreal. *Maybe,* I thought for a brief moment, *this was all a bad dream*. I knew it wasn't. I had gone through a pregnancy. I had seen Jacob on several ultrasounds, but I didn't allow myself to connect with him while I was pregnant because I didn't think I would be around. The reality was that I had given birth and I had yet to meet my son.

I was scared to see him. I was scared of him. His birth had put me in this state, and I was afraid that my apprehensions and fears would impede his spiritual growth. I believe that positive or negative energy can affect people in many ways. If we sense something odd, we might reject it. Living beings can sense things. If you are nervous getting on a horse, the horse can sense it and you will become more nervous. If you are happy and positive, people tend to migrate toward your energy, but if you are negative, even a neutral person who has made no prior judgment of you will be repelled. I was worried that Jacob might not have a natural spiritual connection to me, his mother, that he might reject me.

So many things were going through my mind as my sister wheeled his crib into the room. I looked at him from afar. I wasn't ready to have him brought closer. Jonathan went over to Jacob, picked him up, and brought him next to me. All I could do was stare at him. Jonathan then put Jacob on my chest.

He was breathing so softly and sleeping. He was beautiful, but I couldn't appreciate his beauty. I couldn't wrap my arms around him, with everything I was hooked up to, and I didn't have the strength to try to work around it all. In that moment, I wasn't his mother. I felt like a stranger to him.

Jonathan and my sister were his parents. I couldn't breast-feed him, I couldn't change his diaper, and the worst part was, I didn't want to.

I cried. I told Jonathan to get him off of me, I needed to get some rest. I asked everyone to leave, and Jonathan brought Jacob back to the nursery. Jonathan understood, although he felt a little sad that there hadn't been a bonding moment, especially given all we'd been through together. He was sure it would come later. I wasn't so sure. I felt like a failure. The following day was going to be a big day, so I tried my best to sleep.

I was dirty. I felt disgusting. It felt as if I was surrounded by germs and I could see all of the bacteria around me. I wanted to breathe fresh air and sit in the grass, but that wasn't going to happen anytime soon. I also wanted to pull everything off of me and start running, but I couldn't move my legs. They had swollen to five times their normal size. For perspective, I had weighed in at 173 pounds when I went into labor. I started my pregnancy at 120 pounds. Now I weighed 220 pounds. I had acute renal failure, and the fluids were backing up. I felt like Violet Beauregarde from *Willy Wonka and the Chocolate Factory,* who kept ballooning in size until she almost popped. I didn't know how I could actually expand any more.

I had heard they started me on dialysis while I was in a coma, but what did that mean? I was so drugged up, I was surprised I knew anyone was there. I really don't remember anything from those first few days. At one point I forgot there was a baby in this story and I was just trying to make sense of how I got to this point and how I was going to get out of it. I looked down at my body and saw that there was a catheter inserted to help me pee, there were IV lines running in many directions, and there were bloodied spots on my neck where the port was hooked up to the machine for dialysis. I had another IV piercing my left shoulder. My left arm was more swollen than any other part of my body.

I also had an incision with staples running from my sternum to my pubic bone. The C-section had been left partially open to heal better and to allow blood clots to be pulled out with tweezers by seemingly skittish nurses. At one point, Jane, one of the seasoned nurses, came in and asked if it would be okay to bring in some student nurses to learn how to take care of wounds. These were young girls who had mostly seen healthy births and whose experience hadn't gone beyond keeping small incisions clean. They had seen nothing like this. One of them was helping the resident clean out the incision, and I saw her look like she was about to throw up.

Forget her, I was feeling everything and needed to throw up myself. I had no food in my stomach, so all I could taste was the bitterness of the medicine I was gagging on. I couldn't eat—not that I wanted to in those moments. The edema (the swelling) was so bad that it was putting pressure on my stomach, so even if I had managed to eat a morsel of food, my stomach would have expanded and I wouldn't have been able to relieve the pressure by going to the bathroom. The edema went all the way to the bottoms of my feet, where it felt like constant pins and needles. The intense pressure caused cramping in my feet, which made it very hard to walk. I gave up trying to deal with it.

I also kept telling the nurses the catheter was painful. I complained several times that I felt like I was swelling in that part of my body, but was told that everything was normal. Finally, my doctor came in and saw how swollen my vagina was. They pulled out the catheter. It was instant relief and deflation. Everything relaxed, and in five minutes the swelling was subsiding. Jonathan was again surprised by how well I knew my body, even while I was heavily medicated.

Someone was stupid enough to hand me my phone, and I started making phone calls and posting on Facebook, almost continuing where I had left off. I called my best friend Rosalind to tell her the bris was happening this week. She had no idea what I was saying, because I had a croaky throat and she was unaware that anything had gone wrong. I posted on Facebook that I "dies, but now aliv out coma but liv." When Jonathan started getting frantic calls from friends, he grabbed my phone and saw what I'd posted.

I wanted to be immediately better, immediately. I needed normalcy. I tried to get out of the bed at one point on my own. Big mistake. I watched everyone talking as I "sneakily" inched my way to the edge of the bed, stood up, and then all at once saw everyone again, but at a 90-degree angle. I smacked my head against a pole and the edge of the bed, and everyone screamed.

If I was in pain from that fall, I didn't feel it. The drugs took care of that. Luckily, I didn't pull out any of the lines leading to my vital organs in that instant. I had to promise everyone I would stay put. My type A personality had to be kept in check, but I didn't give in without more fights.

My mother bought me a new dress for the bris, so I could change out of the blood-soaked hospital gowns. I couldn't sit up, and I certainly couldn't take a shower. My sister did her best to comb my matted-down, unwashed hair. Ever try those shampoo bags in a hairnet? I did. They don't work. My face had a yellow pallor, so my sister applied makeup to help me look somewhat alive. I didn't look alive. I actually looked like death warmed over. In fact, my photographer friend Lori, who came to shoot the bris, said there was no life behind my eyes. But I was alive. I was alive, and my husband was looking at me, and I could feel his heartbeat. I could see the tears in his eyes when I could actually look back at him and talk to him. Next to our wedding, it was the happiest I have ever seen him, and lucky for me, my friend captured it on film.

I have no idea how long the bris was, but it was long enough for people in the religious community to come up to me, one by one, to tell me what a miracle it was that I had survived. What a blessing that G-d showed up and bestowed a miracle upon our family. Friends of mine tried to lighten the mood by telling jokes, but I could see the fear behind their eyes as they looked at me in disbelief, amazed that I actually pulled through.

Northwestern is a teaching hospital, so I had to get used to groups from different departments like cardiology, nephrology, pulmonology, and gynecology wanting to visit the "miracle mom." They asked me a ton of questions and then read my charts, and again, one after another kept reiterating what a miracle it was that I had survived. I guess I understood the rarity of surviving an AFE, but their attention made me feel like a sideshow act. I had become a case study at the hospital.

Chapter 8

LIFE IN MY HOSPITAL ROOM became routine. I got used to having my vitals checked at certain times of the day. I also got used to my family having conversations about me in the room, but not including me. The catheter was gone. The tape that had held the drainage tubes in place was removed, along with layers of my skin, causing deep scars. I couldn't lie flat, I couldn't lie on my side, and sitting up was painful. There was no way to get comfortable. Everything hurt, including my psyche.

I was still heavily medicated, but at points of clarity I

began to feel again. I felt intensely sad. Crippling sadness. I was grateful to be alive, but I was bedridden, my brain wasn't functioning, and I couldn't walk. I was afraid I was going to stay that way.

The fear and sadness were usually interrupted by another doctor, nurse, or group of students taking a "field trip." I overheard one doctor say that the hospital had only ever seen ten cases of this rare condition, and only four of us survived. The three others had neurological damage, but my preliminary reports indicated that I was okay in the brain department. They kept saying they couldn't believe I survived. Maybe they thought it would help to hear it, but it didn't.

Why did I pull through when others didn't? It was a question I tried to answer with Dr. Elena Kamel one early morning. She is one of the founders of the Women's Group of Northwestern, where Julie is a partner. Elena had come to remove my sutures with Julie. While she was trying to distract me from the pain, she held my hand and told me she had come to my ICU room every night when no one else was around and brought her siddur, or Jewish prayer book. Elena said she prayed and prayed for me to recover. She told me she noticed my small royal blue and gold siddur on the nightstand and told me that her grandfather had given her the same unique style of prayer book many years before, which she brought with her every night that she prayed in my room. Connected by that prayer book, Elena told me that she had kept close track of my case, reading every chart and talking to the other doctors. With tears streaming down my face and with a shaky voice, I asked her, "Why did I survive?" Elena replied, "I can't give you a medical reason why you survived. I think you need to go spiritual on this one."

Spiritual? Like G-d pulled me through? Maybe. But wasn't it me who had told everyone this was going to happen? How did I know? These "big" questions were too overwhelming to deal with at this time and made me more exhausted thinking about them.

The next day, while hooked up to the dialysis machine for my four-hour "cleansing," another group from cardiology approached and asked to "interview" me. "Do you know how lucky you are to have survived this?" Question after question was hurled at me as if I were being interrogated. It was uncomfortable, and I felt like a science experiment. They talked about me as if I weren't there. I broke down crying.

CARLA, THE NURSE ATTENDING TO MY SESSION, asked me if I prayed and then asked what my intentions were on the outcome of my situation. Maybe the drugs had impaired my thinking, but I wasn't following her in that moment. She then told me a story.

Years before, she had been in a terrible car accident, and she had found herself where I was now—on dialysis to help save her life. She told me that her then-two-year-old daughter kept asking her grandmother why her mommy looked like a monster and that her daughter refused to be held by her. This brought Carla to the brink of letting go and giving in to her misery, but then she made a decision. She said to herself, "I am going to live, and I am going to be stronger than before, and I am not going to let my negativity or anyone else's consume me."

This was not just "positive thinking" on Carla's part, but a strong intention to live. And not just in a physical sense but in a spiritual sense as well. That was it. In that moment, I understood. It was spiritual energy she was telling me to harness to pull myself out of my state of mind. Up until that point, I'd been focused on the physical world. That was the moment when my spiritual recovery started, and it made all the difference in my physical rehabilitation.

That same afternoon my family rabbi from LA, Rabbi Mentz, surprised me by showing up in my hospital room. He told me to change my sadness and frustration with the doctors telling me how lucky I was to be alive into appreciation for the doctors who kept me alive. It wasn't every day, he reminded me, that doctors look at a case and say they have witnessed a miracle beyond all scientific explanation. Their wonder also validated the great work they'd done to save me. I needed to turn this thing around.

Call it mind over matter, but I started getting better. I

was determined. After six days in a coma followed by many weeks of not "being there" mentally, I wanted to feel a part of the world as quickly as I possibly could.

I forced myself to get up and walk. I pushed through the pain, through the blood clots coming out of my incisions and dropping into the already blood-soaked mesh granny panties the hospital provided, and through the enormous weight gain. I pushed through it all.

The constant attention from my doctors and other hospital staff began to wane, and eventually they went back to business as usual. At one point, a social worker came into my room to check on Jacob's status: Could he be discharged now? Jonathan wasn't there, so she asked me to fill out the paperwork on his birth certificate. Still drugged up, I misspelled Jacob's middle name as well as my own. The social worker also wanted to find out whether we had someone at home to help care for him once he was discharged and whether our home would be a safe environment for him. My sister said the conversation between me and the social worker was comical.

SOCIAL WORKER: Do you have help at home?

ME: Oh yes, we have 12 people who live with us and help us 24 hours a day.

SOCIAL WORKER: Okay, that's unique, but great. Do you have any stairs?

ME: Nah, we live in a one-story apartment. [We actually have 32 steps.]

SOCIAL WORKER: Do you have means to get around the city?

ME: Yes, I have a personal chauffeur, who works 24 hours a day for us.

I wasn't being sarcastic. I genuinely believed that. Now, maybe at that point you would think she would stop taking notes and realize I was heavily medicated and slurring my words. But no, she continued.

SOCIAL WORKER: Okay, so driving isn't an issue. Does your husband work?

ME: Oh no, he is retired. He stays at home with me 24 hours a day.

My sister was smart enough to call Jonathan and tell him to get back to the room immediately. He posted a note on the door that said: No one is to talk to Mrs. Arnold without her husband's permission.

Rosalind, my best friend, flew into town to see how I was doing. Picture a laugh-out-loud, hysterical, beautiful, witty Nuyorican—except that she's from Framingham,

Massachusetts—and that's Ro. No one in my family had warned her about the severity of what had just happened to me. Maybe they thought she would be too frightened to come, but they didn't count on the tight bond we share. I knew she would be there no matter what.

Everyone who had seen me over the past two weeks kept telling me how great I looked. I know I looked better than the first day, but the reality was that I didn't feel good and probably looked even worse. And I could count on Rosalind to give it to me straight. When she first saw me, I was being wheeled in from dialysis. I caught the look on her face. Her eyes were filling with tears and her mouth was wide open. She tried the best she could to hide her face behind someone, but I'd seen it. The look on her face told me I had a very long way to go.

My feelings would fluctuate. One minute I was determined to beat whatever this was, and the next minute I was severely depressed, convinced that this was it. I needed to sleep, I wanted to eat, and I wanted to be less swollen and pain-free. I wanted a sign that I was moving forward and wouldn't stay trapped in what I felt was a state of purgatory. It was too massive a request, so I whittled it down to just the next five minutes. I wanted to feel better for the next five minutes.

I asked Ro, "What would you do if you were me in this situation? What do I do, Ro?"

She said, "Ma, I don't know what to do for you other than to pray."

"Okay," I said. "Please let's pray."

Here we were, a Christian and a Jew, asking G-d for help. She said she felt the spirit of G-d in the room while holding my hand, and we started to cry. I felt nothing but an obscene amount of pain and distress, and I couldn't focus on praying. She prayed for me. She prayed that I would get a good night's sleep. She prayed that I would get stronger and that I would look and feel noticeably less swollen.

After I went to bed that night, I actually slept two hours, which felt like a full night's sleep to me. And wouldn't you know it, I woke up a little less swollen and with a little color in my face. When Rosalind saw me, she remarked, "Oh shit, it worked!"

My situation was turning.

The doctors saw improvements in my heart, lungs, blood, and overall physical health over the next few days, but my kidneys were still "off-line." I could go home soon, after they put a semipermanent port in my chest for outpatient dialysis.

Three and a half weeks into my ordeal, I was released from the hospital. I had a slew of doctor's notes, appointments I needed to keep, and medicines and directions to follow to stay on top of my health. It was going to be a long road to recovery, but it was a short distance to home and in that moment that was all that mattered. I was wheeled to the car and got in. I kept looking behind me, like some mistake had been made and I was going to be brought back upstairs. I should have been focusing instead on Jacob's first ride in a car. He would soon see his room, his home, and his new world for the first time.

When we pulled up outside of our front door, I looked up at the sky, I looked down at the rich colors of all the flowers, and then I looked at all the people outside on the street who seemed like they didn't have a care in the world. Their bodies were moving fast, children were laughing, and people were gardening. I began to sob. It was a really strange feeling to be among the living when I felt so dead inside. I felt like the world had gone on without me, which was incredibly unsettling.

Still, I was glad to be home. Climbing the stairs to my bedroom felt like I was climbing Everest without any oxygen. I lay down on my bed and was happy to feel real bedsheets and not smell "hospital." My homecoming signified that I was on the mend.

Then, several hours later, my temperature spiked to 103. My feelings of freedom and hope were dashed. I knew what was going to happen. Jonathan called Julie, and she said the words I was dreading: "Go to the emergency room *now!*"

Chapter 9

Aɴᴅ ꜱᴏ I ᴡᴀꜱ ʙᴀᴄᴋ in the hospital for what turned out to be unsuccessful attempts to figure out what was causing my temperature to rise. One doctor said it was probably a hospital-borne illness. Regardless, the heavy-duty antibiotics worked, and I was released one week after being admitted.

The transition to life back at home was rocky. The nights were when I felt most alone. Everyone in the household would be asleep, but I was too petrified to close my eyes. I was worried that I wouldn't wake up.

One night I found the good-bye letters I had started. As I reread them, with tears in my eyes, I realized that I hadn't finished them because I wasn't supposed to. My signature would have meant "the end," both figuratively and literally. Something had stopped me from finishing them, but I didn't have a clue what that was.

As time went on, the purple-striped incision down the center of my belly started to close, but the C-section was left open. It looked like a second mouth just above my pubic bone. It was extremely painful and looked and felt disgusting. A five-inch window into my innards. The doctors said to keep it moist with saline-filled gauze and to change out the dressing two times a day, removing any clots with a pair of surgical tweezers. Jonathan had learned how to do the procedure at the hospital. With anal-retentive precision, he would line up all the tools and bandages, wash his hands and arms up to his elbows, and put on the special sterile surgical gloves the way the nurses taught him so he wouldn't contaminate any part of them. Then he would flush out the clots, clean the wound, and redress it. He said he liked playing doctor, but I knew that it was more than that. It was his way of showing me unconditional love.

I had never experienced this intense kind of deep-rooted love before I met Jonathan. The love you seemingly only find in movies. My friends and family could see it in the way we looked at each other or held hands every chance we got. We vowed to be there "in sickness and in health," and I guess we got to test that out. Jonathan was passing with flying colors. The catastrophe wasn't changing our bond. It was strengthening it.

The healing had begun. I started physical therapy and began walking again, little by little. I couldn't breast-feed Jacob, and that hurt me more than I can explain. I am not one of those moms who needs to breast-feed until her child is five years old, but with Adina I had enjoyed the bonding, the skin-to-skin contact, and the quiet time. I loved being that lifeline. Jacob was up every couple of hours, and someone other than me would have to feed and change him. I couldn't lift him. I couldn't even feed or change myself. How was I going to take care of a newborn? It was heartbreaking.

If you have ever been on dialysis or had a heart attack, abdominal surgery, or breathing issues, you know that any one of those conditions can incapacitate you for weeks or even months. I had all of them at once, in addition to having a baby and family who needed to see me get better as fast as I could.

Pushing myself, I got better, in relative terms, quickly. After two weeks of outpatient dialysis, my kidneys kicked back in. It felt like I went to bed one night three times my size and woke up the next morning in my skinny jeans. The change didn't really happen overnight, but after seven weeks I finally had the fluid off of me. And because I hadn't been able to eat for those seven weeks, I was skinnier than I had been before I got pregnant. If there was anything good to come out of this ordeal, it was the weight loss. But I wouldn't recommend it as a diet to anyone. Ever.

Even after seven weeks, I couldn't hold Jacob or put Adina to bed. Or go up and down the stairs without getting dizzy. I wasn't allowed to drive. I had major abdominal pain, and I couldn't lift anything that weighed more than a pound. My stepdaughter Valentina had arrived and was a big help with her siblings, but I couldn't even enjoy the little time we had together. I was in a fog. I was, for all intents and purposes, my husband's fourth child. He took care of all of us, and we had no idea how long this was going to last. He had taken an unpaid leave from the New York Attorney General's Office, and we didn't know what we were going to do when it was time for him to return to work.

The same look would come over the faces of all of my visitors when they saw me. They looked like they were seeing the walking dead, or at least that's what it felt like to me. They would resort to small talk: How was I feeling? What did the doctors say? It was all very nice, but nothing too intimate.

My psychologist friend was one of the few people ready to engage and talk about what I had said to her before I delivered the baby. "Of course I was devastated by what happened to you, but I wasn't in shock. You had prepared me for what was about to happen."

Chapter 10

I WAS GETTING PHYSICALLY STRONGER by the day, but every time I tried to start processing what had happened, I would lose it. Especially when it came to my premonitions. How and why had I seen my own death? Was there a deeper meaning to the premonitions? I started searching the Internet hoping to find something . . . anything.

On the first page of my search, I saw story after story about people who had seen their future before it happened. Like 10-year-old Welsh schoolgirl Eryl Mai Jones, who told her mother in October 1966 that she was "not afraid to die.

I shall be with Peter and June." Her mother didn't pay any attention to this strange statement. A few days later, on October 20, Eryl Mai tried to tell her mother again. "Let me tell you about my dream last night. I dreamt I went to school and there was no school there. Something black had come down all over it!" Again, her mother passed off her daughter's comments as just imaginative thoughts. The next day, October 21, a catastrophic collapse at a mine above the town sent a huge, liquefied landslide of water, rock, and shale down the mountain. It destroyed everything in its path . . . including Eryl Mai's school. Eryl Mai, her friends Peter and June, and 113 other children, along with 28 adults, died that day. And Eryl Mai had seen the whole thing before it happened.

Incredible. There it was in black and white. Proof that premonitions are real. I was comforted knowing I wasn't alone.

I kept searching. There was the story of Dan Pearce, a popular blogger who runs a website called Single Dad Laughing. He recalled a time when his son was three and "threw a huge temper tantrum when we were trying to leave the house because he 'didn't want to crash and die.' After about 10 minutes, he just stopped and was ready to go." As they were walking out of the house, a bad car accident happened right in front of them. They would have been involved in that accident had his son not thrown a fit. The child had seen the whole thing before it happened.

There was Trisha Coburn. Her premonitions came to her in a series of dreams in which she was "standing at a barbed-wire fence across from five or six terribly frail people with huge, dark eyes and ghostly pale skin." She felt that they were trying to tell her something. She continued to have the same dream week after week, and each time the number of people she saw grew. And each time she couldn't figure out what they wanted her to know. But Trisha knew something was wrong. She called her doctor and asked for a physical even though she had gone through one six months before. Her blood work came back fine. Another dream brought "100 people wailing, screaming, pleading with me." Trish kept saying, "I don't know what you want from me! Please, please tell me what I'm supposed to do." Trisha's last dream had the same scene, but nobody was there. She pleaded, "Come back, I need you to help me!" A voice said, "Look deeper." Trisha called her doctor and asked what the deepest part in the human body is. Her doctor said, "I suppose it's the colon." She demanded a colonoscopy, and that's when the doctor found a black mass. It was cancer, and it was fast-moving. Her doctor said that if she had waited two more months, she probably would have died.

I could totally relate now. I had seen what was going to happen to me before it happened. And then it had all happened exactly the way I saw it—except for one thing. I didn't die . . . for good. That funeral scene never came to fruition, thankfully.

These cases gave me great relief and confirmation that we can in fact sense things before they happen. That we can "know" things and that we need to speak up about them, no matter how ludicrous they may sound to others—including a loving and supportive husband.

Jonathan read these stories with me but remained unconvinced that any of the events they recounted were spiritually related. I was so frustrated that he didn't even seem to believe these stories, which were so well documented, that I thought, *How can he believe me?* Even though he had lived through my premonitions and then seen them come true, he still had a hard time believing them. He thought there had to be a scientific explanation. His skepticism made me start to question what had happened all over again. I knew I had to dig a little deeper to see if having premonitions followed by the experience of almost dying was just coincidence or if I'd received some sort of real insight into the future. Jonathan's relentless pragmatism pushed me to find answers—because now I was on a mission to prove to my husband that the premonitions were real.

I needed evidence. I needed to talk to some of my doctors who had been there with me through all of the months of foreboding. I wanted to know if there was a scientific explanation for my premonitions. Maybe they'd had other patients who had experienced the same sense of gloom and doom or seen their own "mini-movies." And if there were others, maybe I could glean some sort of pattern or commonality between us. I was looking for any morsel my doctors could give me. Anything.

My first call was to Dr. Grace Lim, the anesthesiologist I spoke to on the phone a month before I died. I asked her why she had flagged my file after I told her about my premonitions, and she said, "It is rare for someone to tell me that they are certain, without a shadow of a doubt, about some bad feeling they have. You were calm but gravely concerned." She said other patients experiencing anxiety about their particular procedure were usually calmed after talking about the risks and possible outcomes. But not me. Grace said my sense of "misgiving was so intense and unwavering" that it stuck with her.

Certainly, the risk of having complications with the placenta previa also caused Grace to flag my file, but she said she was haunted by my last words to her on the phone that day. I had said with a sigh, "It is what it is." Whatever was going to happen to me was already written in the cards, and nothing anyone told me was going to change that or make the fear go away.

Grace said that my sense of foreboding, coupled with the medical risk of the placenta previa turning into something more, motivated her to create a protocol for my case that involved bringing in more blood and additional lines and monitors and having a crash cart available. That protocol, those simple words on a report, helped save my life.

Interestingly, Grace told me about a "feeling" of her own the day I checked into the hospital. She said, "I told Nicole that I was worried something weird was going to happen." It was more than just a fear, she explained. Grace said she knew she had to be in my OR instead of attending to another C-section happening a few doors down. There was really no need for her to be in my surgical room because Nicole Higgins, the other anesthesiologist, was already there. Nevertheless, she felt compelled to be standing there with me. Grace said, "There was something about that day, and I felt G-d's presence and my spirit connected to yours. I needed to be there." Although she considers herself spiritual and believes that "G-d plays a big part in every patient interaction," she said she still can't explain my premonitions, or her own for that matter. "I honestly don't know what to make of our mutual premonitions. It would be difficult to study this scientifically."

Grace also told me that three weeks before my emergency, she and Nicole had gone to the annual American Society of Anesthesiologists conference. They attended a seminar where they listened to abstracts of new studies and medical methods. During that session, a paper on amniotic fluid embolisms was presented, and they learned a new method for saving the life of the mother within seconds of the acute incident. Neither one of them had ever attended to a patient with an AFE before, and there was no reason to believe they might ever see one. "I thought the information was interesting," recalled Nicole, noting that she stored the paper away just as she did with any other interesting abstract. "I didn't think that a few weeks later I would have one." I don't believe that was just a coincidence.

I asked Grace if she was surprised by all that happened. "Of course we were all in shock and taken aback. We were prepared, but you were more prepared. You had G-d's presence all around you that day and probably for as long as the premonitions started to occur."

Early in our conversation, Grace had said that she was more spiritual than many of the doctors she knew. I was thankful for that. There is no doubt she was put into my life for this reason. I also believe her spirit was connected to mine that day and during the weeks before. The OR was prepared to save me that day because of Grace.

I started thinking about that connection and why it happened. Why did Grace randomly answer the phone that day and then instinctively get what I was saying? Why did she have those very real and fearful feelings the moment I was wheeled into the delivery room? Was it coincidence that Grace and Nicole went to an annual conference just three weeks before I died, and that they just happened to attend a seminar where they learned new techniques for saving someone with this rare condition? There was something bigger working here, some giant force putting things in place to ensure my safe return. Nothing about what happened to me was a coincidence. I was sure of it.

I called Nicole and asked her what she thought about my premonitions coming true. "I don't know what to think about it," Nicole said. "As a scientist, it's really hard to explain it. But I can't discount it either."

Lastly, I wanted to talk to Julie, my OB/GYN, to find out her thoughts. After all, she was the one doctor who had been listening to me for months about all of my fears. She had delivered Adina, and she was more than just my doctor—she had become a friend. Still, I had no idea if she was spiritual or religious. I had no idea how she would feel if I even broached the subject, but I was curious to know how she felt about all that had transpired. She readily answered my questions.

"You realize just how spooky this was, that you knew something bigger was going to happen," Julie said. We went back over each premonition, and she confirmed that, except for the one showing my funeral, each one had come true. It was comforting to know that she finally believed me, even if it was after the fact. "It really blows my mind," she said. "I don't know if there's an explanation of how or why you had the premonitions. You just called it, and you were right."

The doctors' responses gave me great relief and validation. They helped me feel like I wasn't crazy. That it had all been real. Here they were, scientists who believed in fact and not conjecture but who seemed to believe that I could see into my future, even though there had been no hard evidence to support it.

It was at this point that I became completely convinced that my insight into what was going to happen to me was absolutely real. Jonathan couldn't argue. Like the doctors, he always explained the world through hard facts, yet all of a sudden the people who thought like he did were convinced otherwise. "It's interesting," he said when I asked what he thought about their responses. Then he quickly changed the subject. It was still beyond his comfort zone.

I was trying to process all of this information both mentally and emotionally, but it was too hard to do alone. I knew I needed professional help.

Chapter 11

I WAS FEELING BETTER, but I also sensed that integrating all that had happened was going to overwhelm me if I did not get some help. Physically I was doing better; mentally I was not. I didn't know how to deal with my emotional pain, and Jonathan wasn't remotely ready to acknowledge his own emotional trauma. In fact, his stoic responses of "I'm good," and "I'm just happy we got through this," were making me feel like I needed to simply appreciate the fact that I was alive and "get over" everything else. I was frustrated that I couldn't just do that.

So I decided my next healing hurdle would be psychotherapy. Therapy had always helped me navigate uncharted waters in the past, guiding me when I was at a crossroads and had choices to make. I knew I needed to discover coping mechanisms for dealing with everything that was beyond my scope to handle. Therapy needed to begin for both of us as soon as possible. Jonathan might not have wanted to admit there were any issues here, but he knew he had to be there for me and go through this with me, even if only as a spectator. I was hoping he would eventually engage and discuss his pain. I was sure it was there, even if he denied it.

I asked several of my doctors for referrals to therapists who had expertise in women's issues, postpartum conditions, and difficult childbirth. I also wanted someone who had the extra credential of having dealt with near-death experiences. It turns out, such a person was not easy to find in Chicago.

My healing agenda was a long one: I wanted to unburden myself of the guilt I felt for not being connected to Jacob initially, I wanted to relieve myself of the survivor's guilt I was feeling, and I needed to face the beast that had tried to destroy me. I needed the courage to read about AFEs, understand the condition, and ultimately let go of my own experience with it. I knew that if I didn't deal with all of this head-on, the fear would bury itself deep inside me and I would never be able to get it out. I wanted psychotherapy to give me the tools I needed to get over the physical and mental pain, but I also wanted someone to help me understand my premonitions and how and why they occurred, as well as someone who could help bring healing energy. These were big shoes to fill.

I met with the first therapist. She was lovely and very accomplished. When we got down to business in the first session, I said, "I need to learn how to protect myself when we embark on this spiritual journey." I'd read that when you delve into any spiritual work, you should visualize something like light or a dome around yourself in order to protect your energy field so nothing can penetrate it while you are learning about the metaphysical. She replied in a very sincere, linear, psychotherapist kind of way, "Why do you feel the need to protect yourself?" I smiled. I knew it would take a lot longer than one session to explain to her what I meant, and I didn't have the time. So I thanked her for seeing me and moved on.

After a few weeks and a few more therapists, I found one I liked. She was in her late thirties or early forties, blond, beautiful, and a mother of three young children. She had a no-nonsense way about her. She wasn't the type of therapist who would say, "What do you think?" She would tell you exactly what she thought and how we were going to deal with this. I appreciated that approach because at this time my "take control" personality had gone into hiding and I needed someone else to take over the reins and just tell me what to do. I trusted her instinctively, I talked to her openly, and most importantly, I listened.

I told her I felt detached from my body, I cried all the time about nothing and everything, and I didn't want to be reminded of what had happened. To my relief, she said that what I was going through was normal. She taught me about acute stress disorder, which comes in many forms.

Jonathan sat with me and listened too. He answered most of the therapist's questions because I kept crying. He took over and asked her what we needed to do. What would be the best way, he asked her, to hit different milestones? How would we know when we were out of the woods? And at what point would she feel the therapy had been successful and we could stop? I thought it was a little too early to ask that last question, but I was proud to have him by my side, being my man, stepping up, taking notes, and helping me get back on track.

She told us that, if after six months I wasn't feeling better, I would move into something more serious: post-traumatic stress disorder. Some cases of PTSD can be crippling, she said. I might be afraid of very simple tasks or even nervous about being responsible for my children. Then a different kind of intense therapy would begin.

That's when I made a decision. I never wanted to teach my children to live in fear, and I needed to lead by example. That therapy helped me gain the strength I needed to face the beast. This thing had almost killed me, and I was not going to let it cripple me for the rest of my life.

Whenever anyone mentioned my experience in the hospital, I would usually deflect conversation about it by saying, "We dodged a bullet." People would accept that and move on to the next subject. I was relieved that we could take the attention away from the 800-pound gorilla in the room in this way. But in one session my therapist said, "You didn't dodge a bullet. The bullet hit you dead-on. You survived the bullet." She told me I needed to accept that, and all of a sudden I had a newfound respect for what had happened. I needed to acknowledge it and give it the respect it deserved.

From that point onward, anytime someone asked me about what had happened, I told them how I survived, not about how I died. That shift opened up conversation instead of shutting it down. I would look the person asking in the eye and be immediately engaged in a deep conversation about how I saw what happened to me before it happened. At times it felt unnatural to be so open with my feelings, but I felt that, if I'd been able to sense that things were going south for me, maybe others could too if they listened to their own instinct. And people listened.

It wasn't about me. It was about helping others understand that if they sensed something, they should say something because speaking up could save their life. Whether it was the barista at Starbucks or a close friend, I spoke to everyone.

One day a friend of mine called to tell me she had a strange pain in her belly and was feeling restless. She had delivered her baby six weeks before. Her husband thought she was being emotional, but she knew something was off. She called to tell me she had an appointment to see her doctor, but she was scared. I asked why she was scared, and she said, "I don't want to die." That put me on high alert. I told her to get to the hospital right away, and I would meet her there after taking care of the kids.

She went in, and her instinct turned out to be right. The exam revealed that they had left a piece of the placenta in her uterus. Removing it, they said, would be a 20-minute procedure. They could do it immediately and all would be fine. No big deal. But by the time I got to the hospital two hours later she was in a medically induced coma and in the ICU. After a few days in the ICU, she came out of it, and when she was recovering at home, she said she would never stay quiet again. She had learned that her instinct was far stronger than she gave it credit for. And her spiritual strength grew exponentially.

My friend's experience was another reminder to me that in our society people aren't always encouraged to listen to their bodies or their feelings. Yet we ignore our deep inner knowings at great peril.

At times I felt consumed with an almost evangelistic zeal to tell my story and encourage other people to trust their feelings, trust their bodies, trust their own inner knowings. If you *sense* something, *say* something. That became my motto, and no one was outside of my reach. As I was telling a plumber about what happened to me, his assistant over-heard me talking about the feeling of foreboding. He asked me what that felt like. Then he told me he had been putting off getting a stress test for his heart even though he sensed something was wrong. He was too scared and thought that what he didn't know wouldn't hurt him. I told him that, on the contrary, hiding from something because of fear could kill you. He called to set up his appointment that very same day. Later he called to tell me that the tests showed some-thing, but he wouldn't elaborate. He did say that he wasn't in immediate danger and doing the test had been the first step in protecting his life. He was grateful for the push.

One by one, people would engage with me about my experience, wanting to know details and how it could help them in their everyday lives. They asked so many questions like: "What were you feeling that made this different from just a passing thought?" . . . "Did you tell the doctors, 'I told you so'?" . . . "Had you felt like this before?" I answered them all, and I always followed up with, "If you sense something, say something."

I was getting stronger and could finally open my computer to read story after story about AFEs and the likelihood of survival. There wasn't much.

I found a story about a couple who had been married for a few years when they decided to start their family. Their first daughter was born and the mom spent several months parenting alone while her husband was deployed overseas. During the birth of their second child, she suffered an AFE. The baby was fine, but mom was not. She lost her life. Her husband would say his military training prepared him to save lives but he couldn't save his wife's. I started to well up.

The next story talked about a very young couple in a very religious community. They welcomed their healthy baby, but when the mom was being brought back to her room to recover, she went into cardiac arrest. They saved her, but the AFE had cut off oxygen to her brain for too long. After many days of being in an unstable condition in ICU, it was determined she would be in a coma forever. Her husband is now taking care of his wife and baby at home, with the help of his community, family, and prayer. Something no husband or father should have to endure, let alone one so young. He was only 25 years old.

By this point in my reading, tears were flowing down my face. There were stories upon stories about AFEs, and the outcomes were incredibly painful to read. There were marathon runners, first-time mothers, young women, older women, black women, white women, Hispanic women,

Asian women, Indian women. I learned that AFEs have nothing to do with age, race, or birth order. What I read confirmed that an AFE is completely random and that the difference between surviving and dying depends on where you are and how quickly it is diagnosed.

The tears were coming at a much faster pace now. Again, the guilt I felt about surviving was becoming a heavier burden, and then I thought about Jonathan and how painful this would have been for him had any of these outcomes been his.

I don't know why, but I kept going.

Another couple was excited to be first-time parents. One day the husband went to work, kissing his wife good-bye as she sat relaxing on the couch four days shy of her due date. When he returned later that morning to check on her, she was still in that same position. Apparently her water broke, she had suffered an AFE, and both she and the unborn baby died right there on the couch. Alone.

Devastating. I was sobbing and couldn't stop for several minutes. When I finally calmed down, I decided I was tired of crying. I looked at myself in the mirror and said, "What right do I have to cry here when these women no longer have the ability to speak?" I knew I had to do something more.

My cousin Sari started researching AFEs and came upon the AFE Foundation and its founder, Miranda Klassen. After Miranda suffered an AFE herself years ago, she felt a calling to start the foundation to help families in the same situation. She helps fathers understand what has happened, directs families to resources, and provides much-needed guidance and support. Most importantly, Miranda is working on research to come up with a way to diagnose the likelihood of an AFE before it happens.

She called Miranda and told her that, when I was ready, I would be reaching out to her.

My cousin stayed with me for weeks after I got out of the hospital, and we would talk into the late hours of the night, philosophizing about what had happened, going through our family history, and discussing what responsibility I had as a survivor. I wasn't completely sure what that responsibility looked like, but Sari was. Although I knew that AFEs are rare, I was hearing more and more about them. Sari, who's a journalist, wondered whether they are rare or just under-reported. She said, "I know you are weak now, but you will get stronger, and you have a voice. You need to use your voice and your experience to help others." Okay. I understood that. I was on a good path and getting stronger.

We felt there was great value in getting the word out about this unpredictable possibility called an amniotic fluid embolism. I did believe that others would benefit from knowing about AFEs and about my premonitions and story of survival. Most importantly, I thought people should know about the AFE Foundation. So, seven months postpartum—

and for that matter postmortem—I agreed to let Sari call a friend at CBS in Chicago, and they picked up my story. My entire catastrophic event was told in just 90 seconds. I was happy I put it out there, but I never expected what would happen next.

Yahoo grabbed it from CBS as its cover story the next morning, and it went viral. Katie Couric tweeted about my story, and so did Victoria Justice. I was "trending." What? I was invited on *The Doctors, Good Morning America,* and *Steve Harvey.* And my cousin was right: AFEs and the foundation were thrown into the spotlight and people wanted to know more.

Days later, when the media frenzy died down, I kept thinking about the last thing Steve Harvey asked me. "What did you see when you died?" I told him, "Nothing." No bright light. No dead relative. I never wanted to make up an answer. The fact is, I remembered nothing. Or at least that's what I thought.

Chapter 12

THE THERAPY I HAD BEEN DOING over the last few months had been good. I was starting to feel like I was among the living again. But I knew I needed to try to remember what happened in the OR that day. When I asked the therapist about returning to the moment I died, she replied, "Why do you feel the need to go back there? You are obviously blocking this out because it was traumatic. Let's focus on something else. You don't need to go there in order to heal."

I was functioning well, healing, and getting back into a

routine. Jonathan was in total agreement with the therapist that I didn't need to remember what happened when I died. He kept asking, "Are you trying to find an answer to help you or others?"

"Both," I replied.

I had lost six days of my life. I wanted to know what had been going on around me and, more importantly, where I was during those 144 hours. My body had been attached to machines in the ICU, but where was I? This question kept nagging at me. There might be something to remember, or there might not. But as long as I had questions that might have answers, I wanted to try to answer them.

Jonathan asked, "What if there is nothing to find out and you go through all of this, opening a wound, just to discover there is nothing there?" I admitted I was scared of the process, but I wasn't worried about the possibility of the therapy not working. All I wanted to do was explore these questions for possible answers. If it led me to a dead end, so be it.

I started looking into nontraditional forms of therapy and came upon regression. Regression therapists use hypnosis to put you into a deep meditative state and take you back to the moment of distress. The goal of this type of psychotherapy is to retrieve memories of traumatic experiences and through insight and emotional release, be able to gain mental, physical, and emotional relief. This sounded promising. I had never heard of this type of therapy before, and neither had Jonathan. I wasn't sure I would be able to relax and be hypnotized, let alone allow someone to guide me. But I knew I had to give it a shot.

Luckily, a friend had told me about a regression therapist she had used before in Miami named Linda Burns. I wanted to do it. Jonathan kept saying, "Why are you doing this? We are good, you are getting stronger, healthier, we don't need to go back." Maybe that was how he felt in the "we" department, but that wasn't how I felt. I was still feeling a heavy burden, and it wasn't coming now from my body or mind. It was coming from something else—from the same place where I had sensed things earlier. I felt like I was somehow being guided to do this.

I explained to him that traditional therapy couldn't help me deal with everything that had happened. "There was nothing traditional about this entire ordeal," I said, "so why would we think the therapy should be?" I wasn't heavy physically any longer, but it felt like something was still weighing me down. I felt I needed to at least attempt to get some closure and to understand what had helped me see the future. Maybe I could do that through regression.

Jonathan wasn't ready to acknowledge that anything other than coincidence and good medicine had saved my life. But he knew I believed otherwise. He wasn't going to stand in the way, but he wasn't going to make it easy for me to share this experience with him. He wanted my desire to do this to just go away as fast as it could.

I was extremely nervous the first time I picked up the phone and called Linda for an appointment, but when I spoke to her I felt immediately at ease. She's a Cuban Jew, just like me, and her slight accent and inflection made her sound like one of my family members. There was great comfort in that. Her voice reminded me of my Grandma Ida. I took that as a sign that this was exactly where I needed to be.

I set up my first session with Linda and recorded it. I videotaped our sessions because I didn't know what I would remember if I was really hypnotized and I wanted to share that information as accurately as possible.

I had never been hypnotized before and wasn't sure I could get to that relaxed of a state. But I was open to whatever came next. Linda explained to me that my mind was my only limitation. It was probably good that I hadn't done much research into hypnotherapy because I would have come to the session with preconceived notions, anticipated too important an outcome, or been disappointed if nothing came of it. I had no expectations, and honestly, I thought I would leave empty-handed.

At this point, Jonathan seemed less concerned about the therapist than about what I would find in the "unknown" places I might visit through hypnotherapy. It was the first time he was thinking from a spiritual perspective, though he probably didn't even realize it.

After getting some background information, Linda told me to relax and try to get into a meditative state. I took a deep breath and closed my eyes. What I saw next blew my mind.

All of a sudden I was whisked back to the OR just before the procedure. I saw my body on the table with the doctors preparing for surgery. My mouth was ajar, but no words were coming out of it. The doctors were "checking in" with me, telling me what they were about to do and asking if I understood, but I wasn't responding.

Linda watched as my face started twisting in pain. I realized I was looking back at the moment when my spirit separated from my body. There I was, standing next to my body, trying to warn everybody in the operating room that I was about to die.

> *ME: I separated because I don't want to see what is about to happen?*
>
> *LINDA: You don't want to see them cut your belly?*
>
> *ME: Yes.*
>
> *LINDA: Do you feel them cut into your belly?*
>
> *ME: No. It's just a body. I don't feel anything. I want to remove myself from being there. I can't do anything. The baby is coming out, but it's too painful, too fearful. So I just separate. I go away while they do their job.*

I began to cry with a feeling of great heartache.

> *ME: I'm waiting for my husband, wanting to hold his hand, wanting to be with him, wanting to be in the labor and delivery room with my daughter. But I don't want to be where I am.*

I felt my chest tightening.

> *LINDA: So, as you start to pull away from that moment, just allow yourself to be there as you separate. What is that like?*

> *ME: I'm doing so much talking to people before going into the operating room, telling them I don't want to be here and I don't want to die. These are the things I was thinking of, these are the premonitions I was having. By the time I get to the operating room, they're about to do surgery, there's no moving, there's no getting out of it. No matter what I say, I can't change the circumstance. . . . I can't move, so I just quiet down and I separate. I just don't want to watch me die.*

> *LINDA: So you choose to separate? You choose to go further away so you don't have to be there? Understandable. And there's a place you go which will come to you. So go to the*

moment where you last remember being right before you
float away, right before you separate.

ME: *I'm freezing cold right now.*

I needed to stop. This was way too much, way too soon. I hadn't been expecting to actually be hypnotized in the very first session. I thought we would talk or maybe learn some breathing exercises. I certainly didn't think I would see what I saw. I was still crying when Linda brought me out of hypnosis. She asked if I was okay, and I answered, shell-shocked, "Yeah, I didn't expect that."

What had just happened? Did I really travel back in time to the operating room? Had I stored all of that experience deep inside my memory and only now was able to access it? Was I making it up, or did it actually happen?

Linda started talking. "This was very real, of course. When I asked you to go back to the time when you separated, did you get a static feeling in your body? Was it kind of . . . it was mostly visual, but did you feel anything?"

"I was feeling a buzzing all around my body." She told me that was a very good sign—a sign that I had reached a deep meditative level.

My fear started to subside, and I began to get excited. I was a sponge, I wanted to understand everything. I explained to Linda more clearly the visuals I was seeing while under hypnosis. I saw my daughter Adina.

LINDA: She was in the room?

ME: No, she was down the hall.

LINDA: So what did you see when you separated?

ME: I was above it [my body], but I was next to it. So I was on my left side. I remember the EKG there. I wasn't on the ground next to me.

LINDA: Did you see anybody in the room?

ME: There was someone standing on my left side that was right next to the curtain. I just don't know who that person was. Dark hair. It could've been Nicole— it could've been the anesthesiologist.

LINDA: What else did you see or feel?

ME: I remember feeling the pressure, the soap going on the skin. I remember all the doctors looking up at the clock, but I also remember only one woman standing to the left side of me.

LINDA: Who was it?

ME: I just wonder if it was my grandmother. Wow.

I was shocked. Was Grandma Ida there? She had been dead for more than 30 years.

It was at that point I became convinced that I had not been alone, both before and during the entire ordeal. My cries for help had been heard—just not by anyone in this world.

ME: Can we take a break for the day? I'm feeling exhausted.

LINDA: I like that you had a catharsis.

I was trying to process what I'd seen and felt. And then I finally understood.

ME: They actually warned me ahead of time. They actually gave me the tools to be able to find the people who would help me prepare for this.

LINDA: You helped save your life. *They* helped save your life.

They had been guiding me. *They* had been around me. I wasn't sure yet who "they" were, but I knew one of them was my grandmother. All at once I felt relieved and scared. And crazy. The logical side of my brain kept rejecting what I'd seen. I asked Linda, "What if it's not real? What if it's all made up in my head?"

Linda wasn't fazed by these questions, which I found

hard to understand. For me they brought up feelings that were extremely hard to reconcile. I also had more questions coming out of this session than I did going into it. What had I gotten myself into?

Drained and disheveled, I said good-bye to Linda.

How was I going to explain any of this to Jonathan?

Chapter 13

ALL OF THE MONTHS of traditional therapy hadn't made me feel as good as this one regression session did. I wanted to tell Jonathan everything, but I didn't know how. It was hard for me to put it all in words, so I just sat him down in front of my computer, pulled up the video, and hit Play. After two minutes of watching, he slammed the computer shut and said he didn't want to see any more. He said he couldn't bear to watch me in pain and reiterated how opposed he was to what I was doing. "You need to stop!" Then he walked away.

I had been hoping for a different reaction, but I guess I wasn't surprised by it. Jonathan works with equations, probabilities, and statistics, and he's an expert witness who testifies in cases where people count on him to give them the facts and to be real and honest. For that reason, there was a major disconnect between my husband's thought process and his attempts to "justify" how and why I had survived. He preferred not to think about it.

Also, the actual event had terrified him, and now the regression therapy was making him even more uncomfortable and scared. In the video, he saw me in great pain, both physically and emotionally, but I also wondered if his reaction came from seeing for the first time what had happened in the operating room. He had not been there in the OR when I died, and now he was watching it as I was reliving it.

He begged me to stop the therapy. He told me, "This is too painful." I understood. It was painful for me too, but I explained to him that I was more fearful of what would happen if I kept all that pain inside me. I didn't want to go through life being scared to drive by the hospital or to deal with any situation that might trigger the acute stress disorder. I wanted to be free of pain, and I was now convinced that this was the way to learn how to deal with it. Jonathan, coming from a family of intellectuals who don't wear their hearts on their sleeves, understood logically what I was saying, but the fear in his eyes said something different. I finally asked him, "Are you scared for me or for you?"

"I don't know," he answered candidly.

Jonathan said he could watch himself experience pain and torture a million times over, but one moment of watching me in pain was unbearable. He continued: "I don't understand any of this spiritual stuff, and I cannot comprehend how remembering these moments could possibly help you. I see it only hurting you."

I got it. I knew that, for now, I needed to move forward without sharing every little detail with him. He wasn't ready to receive it, and I didn't know if he ever would be. But I knew I couldn't stop.

THE FOLLOWING WEEK I headed back into the operating room to have a surgeon fix hernias that had developed in the scar tissue. They call it a Swiss-cheese type of hernia: the scar tissue starts to pull apart, creating holes. The fear is that the intestines could pop through those holes, forcing an emergency surgery. Luckily, my doctors had discovered the hernias before they had progressed that far.

I told Linda I didn't want to do regression as much as I wanted to work on meditation exercises that would keep me calm going back into surgery. I was scared I would have another heart attack during the procedure, and I wanted to try to work through that fear. I also put regression on hold because a doctor friend had told me that it can be harder to come out of anesthesia once you see that "other" world: the search for answers may make you want to remain there in that enlightened state longer than you should. I understood.

As Linda taught me how to breathe into my pain and fear, we talked for hours about intuition. I confided in her about my past intuitive experiences, which I had chalked up to coincidence. I was sure now that they weren't. I knew that, through Linda, I would come to understand more about my intuitions and how in the world I could have felt those things about others before they happened. Most importantly, I would learn how I had been able to see my own future.

I told her about my last conversation with Uncle Marvin and how I'd known I was never going to see him again. I discussed the heart pain I'd felt when Grandma Ida had a heart attack 1,500 miles away from me. She said that many people have a heightened sense of intuition and that intuition is your soul's way of connecting to the soul of another. "Everyone can do it," she explained. "It just takes practice. Yours happened to be gifted to you."

I desperately wanted to know more, and as quickly as possible, but I made a promise to Jonathan that I would not get hypnotized until I was in a safer postsurgery zone.

The surgery day came. After meditating and doing the breathing exercises, I had a feeling of calm unlike the first time around. Jonathan was there, my mother came to town, and my friends showed up. Some of my doctors showed up too, even though they weren't attending at the surgery. I was playing Pharrell Williams' "Happy" and I was ready. Everyone else was nervous, but I looked at them and said, "I don't have any premonitions about this one. It will be a piece of cake." And it was. I went home eight hours later.

My sessions with Linda continued, and I went back into hypnotherapy full force. But this time it wasn't working. I was thinking too hard. Every time I got to the point where the surgery was about to start, I would start hyperventilating and feeling pain in my chest. Linda would need to bring me out. I was blocked, and now I was finally beginning to understand what Jonathan was afraid of.

At one point I was feeling strong stomach pain. Sharp pain. I didn't know if it was the therapy and going back to those moments during the first surgery, or if it was the pain from the recent surgery. I felt a little uncomfortable. I didn't want to go backward with my physical recovery. I knew that going forward with this therapy was the right thing for me to do, but I also knew I had to tread carefully.

Through the next sessions, Linda made me feel safe enough to go back again to that day. We continued with regression as I tried to really comprehend what I was seeing. At the beginning of one session, before going under hypnosis, I asked Linda to help me understand.

ME: It's one thing to see the filmstrip as you explain it, one-dimensional. All of a sudden, through your work, I'm seeing things three-dimensionally. It's so vast. It'll take a lifetime to understand. I guess I just have to accept it.

LINDA: You have to accept what?

ME: I have to accept what those visions were—just that. Because if I dissect it and say, "Okay, how's that possible?" I come back with, "It's an impossibility." I just have to accept it. Blind faith in Hashem [G-d]. I have to accept it. But the miracle that happened to me? Was it a miracle by the hand of G-d? Was it angels that were there to help me?

Linda didn't answer. I think she wanted me to come to my own conclusions. So we moved on to the hypnosis. Maybe I would get clearer answers with my eyes closed.

LINDA: *What do you feel when you separate?*

ME: *I literally feel myself rip out of my body, and I'm standing next to the EKG unit. Next to me, on the other side, is Grace Lim, the only doctor who flagged my file.*

Actually, I explained, I wasn't standing. I was floating a few inches above the floor. Then, amazingly, I floated out of the OR and down the back hallways to see Adina with Tessie in the labor and delivery room. Adina was playing with the blood pressure cuff, and Tessie was trying to get her to sit down and listen to a story. Adina was singing and dancing around and pretending she was Doc McStuffins. It made me laugh, and then I became sad and nervous. I had to go back to the OR to check in on "me" again.

I was hoping that the brutality about to happen to my body was over, but I came back too soon. My listless body, with eyes open, was still on the table just waiting for them to start the operation. I could see that my spirit wasn't planted on the ground. And I could feel it. It felt as if I was as light as a feather. My spirit was actually floating, and I knew my spirit wasn't in my body.

I felt the opposite when I looked at my body on the table. I could feel the heaviness of my body on the operating table as life was getting sucked out of me. My body was just dying. My spirit was standing next to the EKG unit, hearing those last few beats of my heart before I was about to flatline. My spirit was standing in between the machine and Dr. Grace Lim. Grandma Ida was there too. She was rubbing my left thigh, telling me everything was going to be okay.

Linda asked me how it felt to be outside of my body. I told her I felt very light and that I was able to move from one place to another in a flash. She asked me if this was my first out-of-body experience. I told her it was, then said, "I feel like the curtains are being pulled back on Oz." She asked me whether I believed in it.

> ME: Yeah, but there's a twinge of, well, maybe it can be explained by something else.

I was still having a hard time accepting fully what hypnosis was showing me. But I was seeing things I couldn't have otherwise known and getting clarity as to what happened to me in those moments. Linda asked me how I felt about going back in. I told her I was beyond excited, but also feeling afraid to feel the pain again.

> LINDA: I don't know if you need to feel it. If it's too much, you can always disassociate. That means you pull back and you watch it. You are not in your body.

> ME: I don't know how to do that yet.

Even though I had seen myself separate from my body in that early therapy session, I wasn't sure how I could force myself again to separate from what I was seeing. "That's why you're practicing," Linda said. "You're not the first person, you're the second person. It's not happening to you, it's happening to another person." She said to look at what I was seeing like I was watching a movie about someone else. To be a spectator or observer. I closed my eyes.

LINDA: If you haven't done so yet, go to the hospital and be there next to her. So understand, you're there to be of assistance. This has already happened to a part of you. The other part of you is there to help her.

ME: So I'm just there stroking her hand, telling her what everybody's doing. I tell her, "I know you're scared and that Jonathan is not here. But I'm here for you and I hear you. I'm listening to you. It's very real. I'm going to be here the whole time to get you through this."

LINDA: What else do you know about her?

ME: Her heart is beating fast, and the nurse is telling her to calm down . . . that it's not good for the baby, that she doesn't want to put the baby in harm's way, so calm down.

LINDA: So she sees everybody doing what they're doing? What else is going on inside her?

ME: The baby is getting ready. Just moving, kicking. She can't move her arms and her legs. She's shivering. They're asking if she wants more blankets.

I told Linda that I was trying to take her hand in mine.

ME: She can't grab hold of my hand, but she feels the pressure of my hand. She's in tears, telling me that she doesn't want to die. She's angry that she's in this position. She doesn't want to have the baby.

LINDA: She doesn't want to have the baby?

ME: No.

LINDA: Because . . . ?

ME: Because of the way she's feeling. She doesn't want to die, and she's mad for going through the process to get herself pregnant because all of this could have been avoided. She's fearful of that child. She's fearful of that baby because he's going to take her life, and she's also hating herself for feeling that way because this baby is innocent. She wants to love him, but she doesn't.

LINDA: What else happens now? What do you need to do now?

ME: I just have to keep whispering in her ear, telling her what's going to happen, step-by-step, so as she feels pressure, so as she feels iodine going down, she knows what's happening. The anticipation is worse than knowing the moment-by-moment, so I'm there to tell her frame-by-frame what's happening.

LINDA: So you can see what's happening as you stand next to her?

ME: I'm her eyes and ears because she's tuning out. It's hard because no one is paying attention to her, because they're focusing on their job, but they're not paying attention to her real fears. They're just looking at the numbers and just looking at what they need to do.

LINDA: So it must be real hard for people not to pay attention to her. Such a difficult time in her life. . . . So now what happens?

It was surreal to believe I was standing there watching what was about to happen to me. I felt like I wanted to protect my body on the table and stop the doctors from doing the procedure that ended my life. But I couldn't. Linda told me to be the guide, so I started to describe to "myself" on the table what everybody would be doing. I reminded her that she was going to be okay and that this was all she really needed to know to get through this. I promised her, I guaranteed her, that she was going to be fine. No one else in that room could do that for her.

ME: A nurse is checking the tools to make sure it's all in line and everything they need is there. One doctor is telling another doctor some story that I can't quite hear. There's a

lot of hustle and bustle around me. And they're saying:
"Patient: Arnold. Stephanie. Forty-one-year-old
woman. Complete placenta previa. Blood on hand.
O negative." Something about extra IV lines. Heart
rate. Pulse. Give the numbers. I can't hear what they
are. The time is? I don't know. And then the doctors
do a roll call. "Dr. Julie Levitt, obstetric gynecology.
Dr. Nicole Higgins, anesthesiology."

I described how they painted my belly with soap and where everyone was standing. Then I watched as they started the C-section.

> ME: *I don't feel anything. Dr. Higgins is down by my*
> *feet, which is odd because she controls my breathing.*
> *She should be up at the top, not down by the feet,*
> *but for whatever reason she's down there. . . . I see*
> *them pulling apart the fascia, the muscles. I see it,*
> *but there's no hurt. I tell the Stephanie on the table,*
> *"You are going to be fine. You're going to get through*
> *this."*

I watched Jacob coming out, and he was fine, and then, when they went back in for the placenta, I knew what was about to happen. I knew I had to warn myself that I was about to die. I looked up and saw what I can only describe as my spirit with a hand on top of the EKG monitor. *Tap, tap, tap.* Her finger was tapping to the beat of my heartbeat. *Tap, tap, tap.* It was getting louder. It was almost as if she was counting down to the flatline.

> **ME:** *I explain to Stephanie, "It's going to happen very quickly, and I'm going to brace you, and I'm going to hold on to you, and I'm not going to let go."*

The moment was terrifying. I began to cry as I felt that pain all over again. As I was recounting it to Linda, something made me thrust my chest forward as if I was having a hard time breathing. I realized I was feeling the moment of my death. *Tap, tap, tap.* I watched as my spirit's finger slowed down to a final *TAP,* and I heard the beep from the monitor as I flatlined. I looked down at the table and saw my eyes roll back.

> **ME:** *I hear them screaming. They say, "Stephanie, Stephanie, Stephanie!" and Nicole runs to the head of the bed. "She's turning blue," she says, and the EKG machine goes flat and I'm done. My body just collapses. I hear them say, "Hit the button. Call the code. Hit the button."* [I start sobbing.] *They hit the button, and it seems like 40 doctors and nurses rush in and they're, like, "Get the cart." They get it up.*

I had to let go of Stephanie's hand when one of the nurses, who was very big and strong, pushed me out of the way so she could start compressions. Julie kept saying, "This can't be happening, this can't be happening." Nicole said, "It's an AFE, it's the only thing that can do this. She's in cardiac arrest."

I saw a weird flash but was more focused on the chaos, which was at a fever pitch. I watched in horror as they worked on me. It was hard at this point to be just an observer. Not only was I watching what was going on, but I could feel it too. Two different perspectives.

Meanwhile, I noticed that nobody in the OR was talking to me. They were just treating me like a science experiment that had suddenly lost its electrical activity. They were all doing their jobs, but not realizing I wanted to know what was going on.

I watched as Julie stood frozen in shock. I saw Nicole put a tube down my throat and get my breathing back. I gagged. I looked down at my C-section and saw blood pouring out. I felt them jabbing my arm as they put another IV into it and called for another blood bag. They cut into my side to put in another tube. The blood they were putting into me was pouring out of me seconds later. I was feeling lighter and lighter.

I watched every detail as they brought me back to life. I could see my blood on Julie's face as she wiped her forehead, and I could hear Nicole speaking very loudly and asking for different medicines.

It was at that point that I realized my spirit was no longer in the room and I understood what the flash had been. I had seen my spirit shoot up into the sky at the moment of my death.

Fully bonded: After a troubling beginning to our relationship,
I became Jacob's mom in every sense of the word.

Chapter 14

HOW WAS IT POSSIBLE to be back in the OR and see everything? Had Nicole really been at my feet and not where she was supposed to be at the head of my bed? And why did Julie freeze? Was there any validity to what I'd seen?

I kept wrestling with these questions. Maybe it could all be explained as some part of me that could accept it storing the information in my brain when the rest of me couldn't. But then came another question: When I floated down the hallway and watched Adina and Tessie playing, how could I have known what was going on in that room? Was I even right?

As I struggled to process all that had gone on during that regression session, I started thinking about what else I saw in the OR. I'd seen Julie down by my feet, but she wasn't doing the delivery. I'd seen another doctor there cutting into my abdomen and helping to pull Jacob out. Someone who seemed familiar. Strange. And if a curtain had been put up so I wouldn't see the C-section as it was happening, why could I see right through it to my feet and see this other doctor? It didn't make sense.

Was I making all of this up, or did I in fact see the whole thing? The only way to find out for sure was to talk to my doctors and Tessie. But my first stop was Jonathan. I asked him if someone else performed the C-section. He said he was never told that. He said that when Julie came out of the operating room covered in blood, she explained what had happened but said nothing about someone else performing the surgery. He just assumed she did.

I had purposely stayed away from reading my medical records and listening to my doctors' stories about what happened in the operating room. I didn't think I was strong enough yet to read or hear it, but I had another reason as well. I didn't want to know too much before starting regression therapy. I didn't want my recollection to be tainted. I needed to find proof another way.

I sent a message to Nicole, and she agreed to a video-recorded call. I planned to play her the tape from my last session and record her response. I asked Jonathan to join me. He was looking forward to hearing her version of what really happened.

I was nervous. I couldn't bring myself to watch the regression. I almost wanted to just show her one quick moment, not the entire tape, but Jonathan said I needed to show her the entire piece to get her full reaction. I got on the call and played the tape. Jonathan would not watch it, and I had to walk away. I went to the bathroom, broke down, and threw up.

Jonathan came to find me in the bathroom. He picked me up and wiped my tears, and after a little while he said, "Come on, go back and find out what she thinks. I'm here when you need me."

Nicole was silent. She had tears in her eyes. She said, "Watching you, I'm back in the operating room the day that it all happened. You took me right back into that moment. It brings back memories for me. It was hard to watch. I remember what I was feeling in those moments, and you took me right back there. I'm a little skeptical."

My heart sank, but I knew I needed to hear the truth and not what I might have perceived to be the truth. If she told me that none of it was true, I would have closure. I would also know that the therapy was working to alleviate the stress I was under. At least now I would have my answer.

"You are obviously in distress," she continued. "It's hard to watch this. Is it hard for you to watch this?"

I answered, "I had to walk away from it, so yes, it is."

Nicole said that she didn't remember some of what I said, but then some of it she did. "Part of me is a little bit skeptical. The scientific part of me says that you were under anesthesia. You had no blood pressure or circulation. How could you remember this?"

So was she saying that she believed what I was saying in my therapy to be true?

"Part of me is a little freaked out, because it may call into question everything we think about regarding medicine and what we think happens. And then part of me thinks, *wow,* it's pretty amazing."

I didn't want to hear any more. I thought she might change her mind and discount everything. "You know I was under stress too," she said. "Not to the same extent as you were, but it does provoke a little anxiety in me too." She continued: "The skeptical side has to do with a lot I know about medicine, or better, what we think we know about medicine. We didn't put you under anesthesia. You were unconscious with no blood pressure. So I mean, I guess it is conceivable that you could definitely have sensory input from that time. We think about it and we think, okay, if you don't have any blood pressure and there is no blood going to your brain, how are you possibly going to have any kind of sense that was going on? But I mean, I guess you can!"

That was good. What Nicole was saying would give Jonathan a possible scientific reason for some, but not all, of this. I asked her where she had been standing. She told me, "At the foot of the table."

"Why?" I asked.

She said she went to the foot of the table because there were others by my head and she wanted to get a better view. I asked her to tell me if the sensory input she referred to could include sight, because I saw everything happening around me even though my eyes were closed. "Probably not," she said. Still, she had no other explanation. "I just don't know what to make of this."

I screamed for Jonathan after I hung up. Surely he would have to believe this now. He said, "Wow, that's incredible. You sure you didn't hear any of what you saw from her before your session?" Well, now I was getting pissed off.

I immediately called Grace on video chat and asked her a ton of questions. Why had she felt compelled to stay in the OR? How had she felt about me that day? What was she feeling now that time had passed and she'd had time to decompress?

In answering, she was very conservative, showing no real emotion but just telling me that she had felt strongly that she needed to be there in the OR that day. Soon we were in deep conversation, and I was waiting before showing her the tape. I knew the right time would come.

I told her that in my session I'd seen some things in the

operating room. She had been standing next to the EKG unit, I told her, then I explained that my spirit was standing next to her. Then it hit me: my spirit was standing next to Grace, and only Grace, that day because she was the one I felt had been connected to me from day one. She'd known something was going to happen, and so had I. Everyone else was doing their job as if nothing was about to happen. But both Grace and I knew differently. I'd felt safe with her and knew that she would be there in my time of need.

Again, her response was conservative: "Anything is possible. It doesn't freak me out, I am open to it." And then she confirmed that she was in fact standing exactly where I said she was.

In the last five minutes of the call, she explained to me that what happened to me was not what she would describe as a seizure, but more of a gagging, like dry heaving. I decided to show her the tape of the moment I "seized." When she saw it, this straight-faced doctor's look changed. Her lips started to purse, tears welled up in her eyes, and she began to look a little uncomfortable. When the code moment had passed, I turned off the video and asked, "Did it look like that?" She answered, "It looked exactly like that."

The responses of these two doctors only strengthened my growing conviction that what I'd seen through regression was real, but I continued to try to get more validation. I guess I didn't want any loose threads to unravel this new information.

I asked Tessie about the time she spent in the labor and delivery room with Adina. She confirmed that Adina played with the blood pressure cuff and that she had pretended to be Doc McStuffins. How did I know that? she asked. And had I seen her dancing in the room too? "No," I said, "but I must've left before that party had started."

Lastly, I made arrangements to see Julie for lunch. I showed her the tape, and she started to cry. When it was finished, she spoke with both a shaky voice and an unbelievable sense of enlightenment. "It is accurate down to where Nicole was, where I was. Exactly as it happened, you recounted it in detail, and blood pooling up, and I couldn't close you right away and making a drain and waiting in the OR." I asked her if she had said, "This can't be happening, this can't be happening."

"I don't know whether I said that out loud or in my head," she responded, "but I definitely said it." She also corroborated her "frozen" state. She explained that, besides being in shock herself, she'd had to wait for Nicole to finish getting me back up on-line before she could do her job and close me up. So she just stood there. But a nurse had told her she could do something while she was waiting. She could put my uterus back in my body. That had snapped her back to reality.

Then Julie shocked me by going a step further and acknowledging that my description of what happened that day went beyond possible scientific explanation. "There was no way that you would have had any other ability to recount that, in that kind of detail," she said, "unless you were in the spirit and your body was on the table. There was no other way."

Finally, I asked her the question that, more than any other, had become the defining one for me. Had she been the one to perform my C-section? I held my breath.

"No," she replied.

Chapter 15

JULIE'S CONFIRMATION that another doctor had performed my C-section solidified my feeling that I had "been there." I'd seen what happened in the OR. I'd gone back there. But I was still a little confused about whether this was a suppressed memory, the work of my imagination, or some other type of ethereal experience.

I tried to break it down. If I was imagining what I saw, how could the doctors confirm it? If I was gaining access to a suppressed memory, how could I have seen through the curtain blocking my view of my feet and seen that Julie didn't perform the C-section but another doctor did? That left one answer. The experience was an ethereal one.

I didn't need to ask any more questions about what had happened in the OR. I was ready to move on. Jonathan was relieved to hear it, but his relief was short-lived when I wondered aloud, "If I could see the operating room, would it be possible to see more? Was I present during my six days in the coma?" He wasn't happy about that. Back I went to regression.

Slipping into a hypnotized state came easily as Linda gently guided me. It didn't take long. Suddenly I was in my ICU room, watching Jonathan.

> **ME:** *I see him sitting next to me. I see him holding my hand. I see him whispering in my ear, telling me he loves me and that I've got to come through this. He says, "I can't live without you. You have to come through this." He's stroking my hair.*
>
> **LINDA:** *Where are you?*
>
> **ME:** *Over his right shoulder.*
>
> **LINDA:** *You're out of body in the hospital?*
>
> **ME:** *Yeah.*

Over the next few hours, I described how I floated down to the maternity ward to see Jacob. I told Linda I saw my brother-in-law Roy and Jonathan bringing my mother into my room.

> *LINDA: Can you feel her pain?*
>
> *ME: I can feel her pain.*
>
> *LINDA: Do you see what she sees?*
>
> *ME: I do. It's the first time in my life that I actually feel that she feels pain regarding me instead of her own pain and her own stuff. It's genuine pain. She looks like she's going to throw up. She's got a fanny pack around her stomach, and it just looks like it weighs more than her.*

Then, my view changed.

> *ME: Everything is white right now.*
>
> *LINDA: Allow yourself to be in the rays of light. Allow yourself to surrender to this light.*
>
> *ME: It just feels comfortable, warm, pleasant.*
>
> *LINDA: Are you in the light still?*
>
> *ME: It's starting to fade. I'm coming back to the hospital.*

I took a detour and saw my dad, with tears in his eyes, sitting in his kitchen and drinking his espresso.

ME: He's saying the Shema, telling Uncle Marvin to watch over me, and he keeps repeating, "Marvin, don't let anything happen to her, you stay by her side."

LINDA: And as you see this, because you can be in two places at once, can you also be in the light? Do you sense Uncle Marvin?

ME (LAUGHING): Yes. He laughs. He says, "You see, even in heaven I have to take your father's orders. I hear you, Ralph. I always hear you."

I then saw that I was no longer alone.

LINDA: Open up even more . . . to this level where you're at. Tell me what else he says.

ME: Other people are around. They're not threatening, not intruding on the space. They're just curious.

LINDA: What are they doing?

ME: They're looking at me and smiling. I feel like I'm two steps down from them and they're two steps above me. Like I have to climb up to be on their level, but I'm not.

LINDA: Do you want to climb up?

ME: I'll climb up.

LINDA: What else is the significance there at that level?
What do you do there?

ME: Everybody looks normal. They don't look sick, they
don't look like spirits, because I don't feel like a spirit.
There's something about them that's brighter. It's like
after you have a facial and there's a glow on your skin.

LINDA: Can you see their bodies?

ME: Yeah.

LINDA: All the way down?

ME: Yeah, all the way down. I see the body. I see legs.
I don't see feet. And I see the head.

LINDA: What kind of place is this?

ME: Like a congregation. It's outdoors. It's bright white.
It's open. It's got that mistiness under your feet, but it's got
this warmth. It's a perfect temperature. People are dressed
in different outfits, but it's comfortable. Somebody's
dressed in a vest with a button-down shirt with khakis,
and then you have somebody else dressed in a long dress
that's flowing. And you have somebody in shorts and a
T-shirt. Whatever was representative of them at their
utmost comfort or the epitome of who they represented in
a physical form.

LINDA: Can you connect with one or two that are there?

ME: My Aunt Betty.

LINDA: What about your Aunt Betty?

ME: She offers me a Coca-Cola. She's the only one that would drink Coke with me. She was just funny. I can tell that they're watching because she says, "I see your mother is still exercising like an animal. She'll be alive past everybody else we know."

I went on.

ME: They're happy. They're not in pain. They're happy and comfortable, and Aunt Betty looks fantastic. Her skin doesn't look dried out, tired. She looks rested. She looks to be in her late thirties, forties. Not like she looked toward the end of her life. She calls out, "Rae . . . Rachel." My grandmother (on my mother's side) shows up, and she looks beautiful. She says she spent all her time down there counting all her money and "here none of it counts." So they make me laugh.

Then tears started to flow.

ME: I'm sad.

I was crying because I realized I wasn't supposed to be there.

> *ME (CRYING): I love them very much . . . but I don't want to be there.*
>
> *LINDA: Do they give you any message? They see you suffering. They know what you're feeling. Do they have any message for you?*
>
> *ME: My uncle says, "I've always known you to be a fighter. I've always known you to survive." He says, "You know what to do." But I need help, and he says, "You don't need help. You know what to do. You've always known what to do. You knew what to do when you were telling the doctors, you know what to do now."*
>
> *LINDA: They have faith in you. They have faith that you know what to do.*

Suddenly I needed to come out. I felt pain down my throat and my stomach hurt and my throat was dry. As Linda brought me out of hypnosis, I started sobbing uncontrollably. I had seen something that I couldn't explain and had connected to people who had been gone for years. But the true source of my anguish was having actually felt what my body felt while I was going through all the operations.

LINDA: Tell me about this pain if you can. What did you get in touch with? What did you feel?

ME (SOBBING): I don't know if it was the operation. I don't know where I was, but I just felt it everywhere. I felt my whole body with tubes, cutting knives, blood and pulling, and my throat not being able to breathe and my heart hurting.

LINDA: The symptoms should begin to subside. So it was more getting in touch with what happened that day even though you weren't in the body?

ME: Yeah, but I turned around to look at me. When I was standing with my uncle and he was saying, "You've always known what to do," and I turned around, and I saw everything being done. I was watching the scalpel go down my entire abdomen and the doctors use a tool to pull everything apart because of the tissue they needed to get to. And it looked excruciating. And I couldn't breathe because there was a tube down my throat and it was tightened, and my heart was sore from going into cardiac arrest. I realized I was conscious as my spirit looking at me. It was gruesome. It was a horror movie. Put the right music to it, the right lighting, and the right makeup, and you could've scared anybody.

LINDA: That's why you've been staying away from this?

ME: Uh-huh.

LINDA: Even though consciously, physically consciously, you didn't really feel that experience?

That question created more questions. "But how do you know you don't feel it? Because of medication? If the spirit is separate from the body, does that mean you can't feel it? Like, how do you know you can't feel it? Because what I just felt, felt real. How do I know that pain isn't stored someplace and comes out?"

Linda explained that the pain could certainly be stored in some sort of repository inside my brain and my body, but urged me to think of the pain as being in the past. She thought it was time to prepare to do some healing by relaxing and thinking about that bright, white place.

ME: The people, they weren't unhappy, they were joking. They were trying to make me laugh and trying to keep me distracted.

LINDA: What do you make of that?

ME: Probably for the reason that they didn't want me to turn around. They didn't want me to see it.

It was a strange feeling. It was nice to see my uncle, aunt, and grandmother, and I understood that they wanted to protect me by keeping me from turning around. That made me feel safe. But I knew I couldn't stay there with them, and that may have been why I turned around.

> ME: The place I needed to be was with my husband, with my body, so I could come back to be with my husband.
>
> LINDA: Very powerful feelings.
>
> ME: Yes.

Again I asked her: Was it real? Did I really see a congregation of spirits? Linda wouldn't give me a direct answer. She just said that I would know in the future.

I continued to go through hours of regression therapy and to see more things that were inexplicable. During one session, I met my husband's father, Philip, who had passed away long before I married Jonathan. Philip told me about his life's regrets and his feeling that he should have been around more for his family. During the whole "meeting," Philip was playing with a coin that was foreign, larger than a regular-size coin, and made from a metal I couldn't identify. We didn't talk about the coin during our visit, but it was memorable enough that I asked Jonathan about it. He said he didn't know of any coin and was doubtful that I had even seen his father. I asked Jonathan's mother and sister and Philip's brother and sister-in-law about it. No one knew anything about the coin. But weeks later, when I asked Jonathan's brother Jeremy if he had ever seen such a coin that was his father's, he replied, "Funny you should say that. I found a unique coin wedged in the crevice of Dad's old suitcase a while ago." Jeremy said that it was a foreign coin, one he thought was from their trip to Croatia. That made the message clear to me, and the hair on the back of my neck stood up. The message was for Jeremy, not Jonathan. Jeremy had been with his father the moment he died, and he was devastated by it. The message was to try to relieve Jeremy's pain by telling him that there was nothing he could have done to save his father. Philip was only sorry Jeremy had to witness his death and wanted him to remember the happier moments, like the time they had spent in Croatia, not the last time they saw each other.

Now even Jonathan's skepticism was starting to fade.

In the midst of a marathon five-hour regression session, I saw other things and people. I believe I saw my best friend's brother, who died when he was seven, ten years before I knew her. He told me, "Tell Sister I always remember how she twirled my hair. I miss that. And the next time she plays with her son, I will be watching her. I'm always around her." When I called to ask her if that meant anything to her, she broke down in tears and told me she used to twirl his hair all the time. It was something that was calming to both of them. I have no idea where that vision came from or how I saw her brother, but I was learning to roll with it.

It wasn't just visions I was having. I continued to feel things too. I had spasms in my stomach whenever I ventured into the deepest parts of the regression. I felt a tugging in my stomach every time I tried to go back into what looked like a lighted triangle or anytime I saw my spirit shoot up. There was always an attachment to my body. When Linda asked me what it signified, I told her it felt like an umbilical cord tethering me to the real world, "a physical reminder," I said, "that we are connected. And if I let go or become detached, all I have left is my body and I would be dead."

That connection was a lifeline to the afterlife, and all of a sudden it was pulling me back to this life.

Chapter 16

AFTER THAT REGRESSION SESSION, I realized I had to stop living in the past. My journey from death back to life had taken both a physical and mental toll on me, but I was determined to live life to its fullest now. I had an amazing family who needed my complete attention, and a loving husband who needed me to be whole again. I wanted to open myself up to this new life, but I needed to have complete closure first. I decided that would come on Jacob's first birthday. My re-birth day.

I wanted . . . needed . . . to thank everyone who had

worked so tirelessly to help our family stay a family. And I needed to go back to the hospital and thank everyone who had helped save my life.

As we drove up to Prentice Women's Hospital, a wave of nausea swept over me. I kept reminding myself that I was there only to visit and not to stay, and the nausea went away. I took a deep breath as I got off the elevator on the eighth floor, where I gave birth. The blood drained from my face and the sick feeling returned. Jonathan held my hand tighter and told me, "It's okay. I'm here . . . this time."

We continued on our way as I tried to gain more confidence. We had arranged for the hospital to gather as many people as possible who were there one year before, and we were escorted upstairs to the nurses' station. Our chaperone guided us toward the room, but I didn't need her to show me the way. I knew where it was. I turned to the RESTRICTED PERSONNEL ONLY door and asked if we were going that way. I knew that once I went through those doors, it would be the third hallway on the left. I knew all of this because it was the hall I had floated down to watch Adina and Tessie playing. I had never been back through there at any other time. It felt like déjà vu.

As we entered the room, nurses started coming up to me, hugging me, shaking my hand, and telling me they were on duty the day I gave birth and coded. Some had been in the delivery room and kept everything running. They might not have performed starring roles, but their parts had been essential to my survival. I knew that. They said they were thankful to see me in such good shape. I told them I was the one who was thankful.

One nurse introduced herself to me as Jessica. She said, "You probably don't remember me, but . . ." I interrupted her and said, "You were the one who broke my ribs." I knew immediately that she was the nurse with the strong hands who pushed the "observer" me to get out of the way when she started chest compressions. I had seen her through regression therapy. She smiled and said, "I would break your ribs all over again because it helped save your life." With tears in my eyes, I hugged her and told her I would be forever grateful.

We handed out cards and small gifts to everyone on the floor who had helped me that day and served as lifelines for my family throughout the ordeal. It was a small token compared to the gift they had given me. Many of them told me they usually didn't get to see patients after they left the hospital. I was glad to be able to show them how their incredible efforts paid off.

I went down to dialysis. I saw the same two patients who had been lying next to me when I was going through treatment there many months before. They were still there, getting dialysis for their kidney issues, and I felt great empathy. Then I got hit with another wave of nausea. Jonathan held on to my hand, instinctively knowing what I was thinking as the head nurse inside the room asked if she could help us. I asked to see Carla, and the woman looked at me and said, "You look familiar."

I said I had been a patient there the previous year, and I wanted to say thank you to Carla for helping me. She said, "Oh my, you were our patient, I didn't recognize you. You look wonderful." At that moment, Carla stepped out from behind the curtain as she was getting done with another patient. We hugged and both of us started to cry.

As we were wrapping up, a doctor came up to me and introduced herself. Dr. Hyo Park said that she was in the operating room that day. In fact, she said, she was the one who delivered Jacob. There she was, the missing link. The face behind the "familiar" doctor at my feet I had seen in regression. She went on to tell me that she had met me before that day. She was the resident in Dr. Schink's office, the gynecologic oncologist I went to see when I was having my premonitions. Dr. Park was the resident whose name I never got and who had sat there without saying a word during that office visit, just taking notes. She was the "other" doctor Julie had told me about. Dr. Park told me she also helped care for my wounds after the surgery. She was there before, during, and after. Wow. It wasn't a coincidence.

I asked her at what point she realized that what had happened to me was what I had predicted. She said, "I was in the operating room when you coded, but I didn't know it was you. I was in the second operation hours later with Dr. Schink, but I still didn't know it was you. They sent me down to speak to your husband after we had you stabilized in the ICU, and as I was walking to your room I caught sight of your husband, and it stopped me dead in my tracks. My stomach dropped and I said, 'Oh my G-d!' "

"Where do you believe the premonitions came from?" I asked her.

"I honestly don't know," she replied. "I know we have only learned about 10 percent of the brain, and what we don't know or cannot answer might be within the other 90 percent." It was the scientific answer I expected. The next thing she said I didn't.

"It was a little freaky and still freaks me out thinking about it. It's nothing they prepare you for in school at all. But personally, I think those visions helped save your life."

The last piece of the puzzle was complete. I finally understood that not only was I saved to be around for my beautiful family but that I also had a mission. I needed to pass along the message that if you sense something, say something. Doing that had saved my life, and it could save someone else's.

A friend of mine suggested that I put together a website that could serve as a resource for other AFE survivors and their families. I created www.stephaniearnold.net and started doing tweet chats and blog posts, talking about everything from pregnancy complications to love. The reaction was overwhelming, and it was continued therapy for me.

> Your story is inspirational and my awareness of AFE moms, families, and their needs is forever growing. Sharing your story and the many stories of others is how I plan to help your efforts of healing!

> Stephanie, You are a modern-day miracle who is now able to educate women and the medical community in much greater detail on AFE.

> Your experience prior to giving birth will also be able to help women become more in touch with their instincts and can teach them to trust their feelings.

I touched base with Miranda Klassen at the AFE Foundation, and she asked me to sit on its board. The coverage of my story had thrust AFEs into the global spotlight, and people were starting to pay attention. Through social media, people found me, tweeted my story, friended me, and commented from as far away as New Zealand and Australia.

I told people wanting more information to search for the AFE Foundation's work with Baylor College of Medicine and how it had created the first international registry for AFE research. Baylor started receiving more inquiries and requests for patient information. Now that more people were learning about AFE and AFE research, more cases and vital information would be collected. Ultimately, the spread of information should help answer questions that have been plaguing doctors and families for years. I felt really good about this work. It was good—really good.

You know how when you are getting married all you see are wedding magazines and wedding shows you never noticed before? That's how I was feeling about AFE. Maybe it had always been this way, but every week, all of sudden, I heard about another case, another tragedy, another family in crisis. Women were reaching out to me and asking me about their sense of foreboding and what they should do. I would direct them to the foundation and give them questions to ask their doctors. I was amazed that people were seeking advice from me.

I received a call from the magazine *Today's Chicago Woman*, which wanted to include me in its "100 Women of Inspiration" issue. It was an honor to be among truly inspirational women, even if I felt like I paled in comparison. All I did was die and come back to life. These women were helping to save lives. That's truly inspirational. I needed to take what I had learned from my experience and do the same.

MY AFE EXPERIENCE changed me and the way I relate to people. But I wasn't the only one affected by it.

My doctors say that my AFE experience has changed the way they deal with patients. Nicole says she now realizes that her interaction with patients, albeit brief, is not inconsequential. "If a patient voices certain concerns, I never discount them now."

Julie told me that her eyes and ears were opened especially after watching my regression tapes. She was saddened by just how alone I felt. Now, she says, she won't lose sight of the "soul," even in an emergency or code situation. She believes the soul is watching, the soul has feeling, and the soul can be talked to. She wishes she'd had that insight when I coded, but she knows now and says she won't ever forget it.

My AFE has become part of me. Even if I never learn why it happened in the first place, I have accepted what happened and how I got back.

Chapter 17

I WAS NOW STRONG ENOUGH to look through my medical records. I wanted to review them and send them to Baylor researchers so they could add me to the registry. I thought maybe something would jump out at me, something that I had considered insignificant at the time but could be important.

I read how the "patient is intubated, not responding, given Halcion because patient is agitated." I read how the doctors would "see how patient does over the next few hours and if patient will be stable enough to continue with another surgery." I read the details of how my body was hemorrhaging and the details about how many doctors and nurses clocked in to keep me ticking.

I was okay. I wasn't getting sick from reading it, so I kept going.

I read about the clamps used to pull my abdomen apart and the drainage tubes and some things I didn't know, like the fact that my bladder had collapsed. I looked at my cousin Sari, who was reading the records with me, and said, "What? How did I not know that?"

There were pages and pages listing so many medicines and updates it made my head spin. There were more than 600 pages in the stack of papers. We made bets on how much the entire hospital stay had cost, and I was closest at $900,000. The records showed my ICU bill broken down by meds, staff, and room charges.

I was happy to realize that the therapy must have been working: I could finally read my files not as though it had happened to someone else but as it happened to me. And I could no longer feel it. There was only one statement from a doctor that crippled me, bringing me to my knees with heart-wrenching tears: "Patient's husband has been made aware of the grim outlook." More than any pain I had felt for myself, the thought of what Jonathan had gone through cut me wide open. It's hard to explain, but we had been spiritually connected from the first time we met. I could feel his pain as if it were my own.

I wondered whether I should go back to Linda for another regression session to see whether I could alleviate his pain now that mine seemed to be under control. It sounded logical, so I called her and set up the appointment.

What I thought would be another regression session turned into more of an analysis of what I'd been through with regression and why I needed Jonathan to be on board with what I saw. Linda started out the session telling me that in all the years of doing this type of therapy, my case was very unusual.

Typically, Linda said what people see during an out-of-body experience is the lifeless body from up above. My experience was different in a number of ways. Under hypnosis I was able to experience what my physical body felt as well as what my spiritual body experienced. I also created a safe space to help heal the part of me that was helpless and about to die.

ME: Does that mean that it was inaccurate? Do you question it?

LINDA: Not at all. There are so many things we don't understand. In doing this work, I've heard people's experiences that are unbelievable. But it's the healing that happens afterward that speaks very loudly. Something happened. In your case, we were able to verify it. In most cases, I try to verify what the person has felt or seen, but usually, [the most important] verification is the patient heals.

ME: As this has gone on, we have done a lot of sessions. At any point did you say, "This isn't real. I'll let her take me down this path. Maybe this is in the back of her mind." Was there any time that you felt that way?

LINDA: I don't judge, but it was really important for me that you told me you had verified it with two of your physicians who were present in the OR. And I heard their comments. They were very detailed comments, not only of what you heard but of what you saw.

ME: What does that do for you as a therapist, coming out the other side of this with me and seeing how I am getting better?

LINDA: It's verification that people have abilities that we really haven't studied completely. And that if you are at least able to be open-minded, even though it may be something different for the therapist, or for me, sometimes amazing things come out of it that are almost miracles. [For instance], releasing, relieving the person of pain, sometimes of physical symptoms. Yeah, so when you do this kind of work, it never ceases to amaze you. At least it doesn't for me.

I told Linda I was 100 percent sure that the premonitions came from the spiritual beings who had been surrounding me. They were one and the same. I explained to her that I desperately wanted Jonathan to understand what I had seen and felt. He looked at everything that had happened as evidence of my own good instincts, even though I kept telling him the premonitions came from someplace else. Linda told me that I really didn't need him to understand it, only that it was more important for him to be there for me. I didn't agree.

Jonathan witnessed everything—he saw reactions and heard testimony that contradicted what he said he believed. He heard me relay messages from the "other" world that were accurate. He went back and read my Facebook postings and good-bye letters describing ahead of time everything that would happen to me. He heard scientists admit that they had no explanation for what happened. And through it all, he had remained skeptical. He was the single most important person in my life. His opinion was the only one that mattered to me. I knew Jonathan better than that. I knew he believed, I just needed to hear him say it.

I left the session understanding more about regression therapy but, sadly, not having figured out a way to alleviate Jonathan's pain, which I knew he was suppressing.

A few weeks later, I was cleaning out our office and came upon a small prayer book. Inside was a sign that Jonathan had made while I was in the hospital and affixed to the door of my room: CANDY FOR MEDICAL ATTENTION. I smiled. On another sheet of paper he had calculated the probabilities of my being diagnosed with this rare condition. And inside the back cover was a third piece of paper. I opened it and tears welled up in my eyes as I read it.

May 31, 2013

My only love of my life, I am sitting here in the still of the night listening to the machines beep around you. You look like an angel, which devastates me. What am I to do if you are gone? How will I ever reconcile this to our children, to myself? I should have listened more, I should have been here, if for nothing else, but to hold your hand. I should have prayed, for you. Rabbi told me this is a test of my faith. If I was being tested, why isn't this happening directly to me? Why to you? What did all of this mean? How did you know? I can't lose you. I would be a shell of a man if you die. Please don't die. I don't

He never finished this letter.

That night I explained to Jonathan that I understood his pain and how he was protecting himself from feeling it. I showed him the letter he had written, as a reminder of his vulnerability, and I told him I needed him to watch a portion of one regression tape. Once again he pushed back.

Then I told him I needed him to watch it not only for him but for me too.

I set the computer down in front of him and cued up the portion of the regression session where my spirit was watching him in the ICU. On the tape, I said I saw him holding my hand, whispering in my ear, and wondering how to heal me. He watched until the end without saying a word. When it was done, one single tear fell down his face. I said nothing. Instead, I picked up my phone and texted him some questions.

"Do you believe I saw your father?"

"Do you believe my premonitions came from somewhere outside of science?"

"Do you believe I saw others?"

His phone lit up, he read my questions, and he started to type. I thought I was going to get another long-winded answer, questioning the probabilities and looking for answers with no conclusions. I was wrong.

"Yes."

"Yes."

"Yes."

Then he added three very impactful words as the tears streamed down his face:

"I believe you."

Chapter 18

A FEW MONTHS AGO, I had a dream about my Uncle Marvin. I remember waking up and thinking he was in the room. And even before I got out of bed, I smelled it. Cigarettes. I no longer had a craving for one, as I did throughout my entire pregnancy. I now believe that craving was a connection to my uncle. It was his way of letting me know that he had been with me the entire time.

When I started this book, I knew I wanted to have 18 chapters. The number is significant in the Jewish religion, as it means "life" (or, in Hebrew, *chai*). Interestingly, my husband took only three pictures of me at the hospital during my entire stay there. One in particular stands out. I am in the coma, in my bed, hooked up to 10 IVs and many machines with tubes running everywhere. In the lower right-hand corner is the machine that kept my heart and lungs going. On the side of that machine, emblazoned in big, black numbers, was the number 18. *Chai.* Life. The numbers were already telling the story. I was going to live.

I am not the same person I was before all of this happened. I love more deeply than I did before. I care more about the little things. I celebrate my new insight and believe that everyone should listen to themselves no matter how crazy they think they may sound. If you sense something, say something. The forces that saved my life have blessed me. I no longer wonder whether my premonitions were the work of divine intervention. I know they were. I no longer question my faith in the spiritual world. From my firsthand experience, I can say that it unquestionably exists.

My therapy is far from over, but each time I go back I go forward in life. I wholeheartedly believe that we can foresee the future and that we can travel back to see the past. My journey took me to places I never thought I would visit and helped me come to a different perspective on what happens when this life is over.

I was recently at a funeral for someone who was beloved by my beloved. She was the mom of my husband's childhood friend, and she had treated Jonathan like a son. She was well respected, loved, and admired by many people, including high-profile politicians and celebrities. I hadn't known her well, but I knew her enough to like her and to know that she cared about our family.

I had been to funerals before. It is an unfortunate part of life. It usually marks the end of a person's life and the beginning of the pain for the family left behind. There's great sadness surrounding death, and this funeral was no different. I felt great sympathy for the family, but I had no tears. Instead, I found myself smiling for this body we had just lost. I say "body" because that is all it was. It wasn't her spirit, or her life, that had died. Only her body had died, and we were saying good-bye to what had been her physical representation.

I smiled because throughout my regression I had seen places and people I had never thought I would be able to see when I had my near-death experience. I hadn't really known this sort of place existed. I had seen those places through my spirit's eyes, and I knew that day that this friend would be with the others I had seen in that airy, beautiful, open space. I could visualize where she was and how fabulous and young she looked. So while her family and friends in the physical world said good-bye, I knew that her late husband and son were welcoming her on the "other side" with open arms.

So yes, I smiled, hoping I wouldn't be seen by Donald

Rumsfeld, just one of the VIPs at her funeral. I smiled as I thought about this feisty woman who loved to hop on an airplane and fly somewhere at a moment's notice. She had always been up for any adventure and couldn't sit still because she needed to conquer the world—this world where we stood at her funeral was far too limiting for her with her physical pain and her tired body. But where she was headed, her spirit would soar and the boundaries would be limitless.

My journey back from death and the events that followed changed me. I guess T. S. Eliot said it best: "It is worth dying to find out what life is."

Just a "normal" family: A year after our ordeal, Jonathan and I are very aware of how precious our time together as a family is.

LORI ALLEN PHOTOGRAPHY

About the Authors

Stephanie Arnold, in her "former life," was a TV producer who spent twenty-seven years creating and directing TV shows, music videos, and documentaries. She left the "business" in 2008, after meeting the love of her life. From that point on, the only thing she wanted to produce was a family.

It was during the birth of her second child that Stephanie suffered a rare and often fatal condition called an amniotic fluid embolism (AFE) and died on the operating table for *37 Seconds*. Everything she does now is a direct result of her survival.

Stephanie currently serves on the board of directors for the AFE Foundation, speaks on patient advocacy to organizations like the American Society of Anesthesiologists, and has raised money for Northwestern Memorial Hospital's Prentice

Women's Hospital. Her experiences led her to be named one of *Today's Chicago Woman*'s "100 Women of Inspiration." She has created the website www.stephaniearnold.net, where she blogs and offers support to families who have been touched by AFEs. Stephanie seeks to instill her message wherever she can: If you SENSE something, SAY something.

Stephanie lives in Chicago with her husband, Jonathan, and is the loving mother of Adina, Jacob, and stepdaughter Valentina.

Sari Padorr is an award-winning journalist whose TV career has spanned 29 years as a reporter, anchor, host, producer, and news director. She is the recipient of two Emmy Awards and a Gracie Allen Award. Her articles have been published in *The Denver Post* and several magazines. She lives in Florida with her husband and daughter and is Stephanie Arnold's cousin.

Acknowledgments

Many people have been instrumental in making this book a reality.

Mel Berger—This book exists because of you (couldn't do it in five words). *Besheret.*

Mark Itkin—Thank you for your open heart, deep respect, and for connecting me to the man who changed my path.

Mickey Maudlin—It took a publisher with gut instinct to get that this book went far beyond 37 seconds of dying.

Dr. Elena Kamel—You are a rock star to some, but you are my rock. Thank you for the 2 A.M. visits each night in ICU, when no one knew you were there. Or so you thought. I have so much love for you.

Dr. Julie Levitt—"Thank you" doesn't quite cut it when

we go through something like this together. You are my hero, my savior, and you have been my entrée into understanding doctors are people too. Thank you for taking really great care of me. You will always be in my heart.

Dr. Nicole Higgins—I know you never want to hear me say "thank you" again, but I don't care. Thank you. We are the same age. We could have grown up together, but we would have never met except under these circumstances. Without your quick thinking, this ending would have been quite different.

Dr. Grace Lim—You are an amazing doctor, human being, and spirit. Your residents are lucky to have you as their guide. Never lose the spiritual connectedness you have and NEVER change the way you practice medicine.

Dr. Hyo Park—You were "just" the resident. But somehow you appeared at all key points in the course of my medical care, conducted the key procedures, and played a crucial part in my recovery.

Linda—Without you I would not have come so far so fast. I am grateful for our journey together and this is not the last time you will be seeing me. You are a gift to this world and the world beyond that which most people see.

To my friends and family, Andras, Divya, Camille, Joy, Mark, Steve, Chen, Lori, Gina (Hope for Accreta Foundation), Rabbi Mentz, Andy and Michelle, David and Melissa, Gael, Jeremy, Alice, Catalina, Mindy, Jessica, Jodi, Rachael,

BJ, Robyn, Sheila, and Rosalind—We all know it takes a village and there is no doubt that this village helped me heal faster, focus on the important things in life, and re-energize our lives. Thank you.

Northwestern Memorial Hospital—Kim Armour, Nick Rave, Dr. Kyle Mueller, Dr. Julius Few, Dr. Regina Stein, Carla, Jane, and all of the support staff from the different departments that heal the brain, body, and soul. Thank you for all your care and concern in helping me to survive. I will always look to your hospital as the best in the world. Thank you.

Mom—Your strength and your determination got me to where I am today. Thank you for picking up the pieces, cleaning up the mess, and making sure I would feel protected.

Papi—Thank you for your prayers and your love. You always told me I would know when I found love; that it would hit me over the head—and you were right. You have always been my rock and I am proud to be your daughter.

Michelle—You are an amazing sister. I know you would have moved mountains to come between me and "that moment." You have your own amazing senses—and you act on them. Never doubt that connection you have inside of you. We are linked on more than one dimension and this proved it. I love you.

Roy—Thank you for being that shoulder Jonathan needed. He was able to be stronger because you were there.

Mark—You are one of my biggest supporters. Your spiritual growth and guidance have helped me wade through these unchartered waters much more smoothly than I would have without your help.

Rachelle and Carolina—I cannot express the love I have for you both. Thank you for loving me the way you do. Alan and Jeremy, you are both very lucky.

Addie—Thank you for letting me borrow your mother to write this book. You have amazing parents.

Tessie—Woman. Not a day goes by that I don't thank G-d for you coming into my life. You have been there for me when there was no way I could show up. You have been there for Jonathan when he didn't have the energy to talk and you have been there for the most important little people in my life. G-d allowed you to walk into my life and no matter where you are or where you go, know that I am always there for you. Always.

Barri—Thank you for believing I could do this.

Attorney General Eric Schneiderman, Executive Deputy Attorney General Karla Sanchez, and the attorneys and staff in the New York Attorney General's office—My husband loved his position and serving the people of New York. I apologize for derailing his government service early with this awful experience. Thank you all for your patience and understanding.

American Society of Anesthesiologists—Thank you for being the first to tell my story and for encouraging me to keep telling it. No one knows how valuable physician anesthesiologists are until they desperately need one. Your doctors save lives. *When Seconds Count*—you want them to be there. https://www.asahq.org/whensecondscount.

Miranda Klassen—Thank G-d you survived. Without you, thousands of people would be lost and the information about AFEs would be virtually inaccessible to laypeople, victims, and their spouses. You are more than an AFE survivor. You are more than my friend and colleague. You are my sister. For the most up-to-date information on AFEs and how to help, please log onto www.afesupport.org.

Sari Padorr—Well, cousin, if you didn't make me tell this story to CBS Chicago, where would I be? I think your intuition is spot-on. I think you are an amazing writer and have a great instinct when it comes to telling a story. Without you, I would have been frozen in many parts of the book. Without you, I would have missed elements from our childhood that were pertinent to what was happening with me in therapy. And without you, this book would not have been what it is. You remember your promise to me. Thank you for flinging open the doors to Caribou Cabin and making it a writer's haven. I love you like my sister.

And finally, I'd like to thank the force and forces that came together to ensure my safe return. I don't just thank you; my family is forever grateful.

L'chaim! (To life!)